ANALYSIS AND CONTROL OF FLOWS IN PRESSURIZED HYDRAULIC NETWORKS

Analysis and Control of Flows in Pressurized Hydraulic Networks

DISSERTATION

Submitted in fulfillment of the requirements of
the Board for Doctorates of University of Liège
and of the Academic Board of the UNESCO-IHE Institute for Water Education
for the Degree of DOCTOR
to be defended in public
on Wednesday, April 26, 2006 at 15:00 hours
in Delft, the Netherlands

by

Rakesh Kumar Gupta
born in Lakhimpurkheri, India

Master of Technology, IIT, Delhi, India
Master of Science, UNESCO-IHE, Delft, The Netherlands

Taylor & Francis
Taylor & Francis Group

LONDON AND NEW YORK

This dissertation has been approved by the promoter
Prof. B. Petry   UNESCO-IHE Delft, The Netherlands

and co-promoter
Prof. dr. A.G.H. Lejeune   University of Liège, Belgium

Members of the Awarding Committee:

Chairman   Director UNESCO-IHE, The Netherlands
Prof. B. Petry   UNESCO-IHE Delft, The Netherlands
Prof. dr. K. Vairavamoorthy   UNESCO-IHE Delft, The Netherlands
Prof. dr. A.G.H. Lejeune   University of Liège, Belgium
Prof. dr. M. Pirotton   University of Liège, Belgium
Prof. dr. J.C.P. Liou   University of Idaho, USA
Prof. dr. H.B. Horlacher   University of Dresden, Germany

Published by Taylor & Francis
2 Park Square, Milton Park, Abingdon, Oxon, OX14 4RN
270 Madison Ave, New York NY 10016

Transferred to Digital Printing 2007

Published by Taylor & Francis Publishers, a member of Taylor & Francis Group plc
www.tandf.co.uk

ISBN 0 415 40917 9 (Taylor & Francis Group)

Acknowledgements

During the course of my PhD, I received help and support from a number of scholars, friends and institutions. Here, I would like to record my deep sense of gratitude to all those who have helped me all along.

First and foremost, I would like to express my sincere gratitude to Professor Bela Petry, my promoter, for his meticulous guidance, valuable suggestions, thought provoking discussions and insightful comments on my thesis. He furnished a fertile environment in which the ideas presented in the present study have developed. His profound and extensive learning, keen perceptibility and rigorous scholarship are absolutely admirable. I am also thankful for his constant encouragement, inspiration and support, for the amount of time he has generously spent in advising me and for his friendly assistance in all aspects of my study and living in Delft. It has been a great pleasure and exciting challenge to study and work under him. His enthusiasm in academic research, thorough knowledge over a wide range of speciality subjects, initiative in educational development, expert teaching skills and profound friendship with students have impressed me and set an example to me.

I take this opportunity to express my sincere and whole-hearted gratitude towards my co-promoter Professor André Lejeune (University of Liège) for his help and intimate co-operation at every juncture of this thesis, without which this work would not have got this shape. I was greatly encouraged and helped all along by him, and I do not suppose I can ever forget his kindness and help. I am thankful for his interest in my PhD work as well as for those marvellous courses he offered during my MSc studies, which expanded my knowledge and consolidated my background.

I express special thanks to Professor Jim C. P. Liou (University of Idaho) with whom I interacted during my PhD study. With his vast practical experience in the field of my thesis work, he offered lots of practical suggestions and guidance. I am thankful to him for taking interest in my work, for his valuable suggestions and for serving on my examining committee.

I would like to thank the members of the PhD awarding committee, in particular, Prof. Richard A. Meganck (UNESCO-IHE), Prof. M. Pirotton (University of Liège), Prof. K. Vairavamoorthy (UNESCO-IHE) and Prof. H. B. Horlacher (University of Dresden) for showing interest in my work and I gratefully acknowledge the valuable recommendations provided by them.

I consider myself fortunate to have been associated with the dynamic and friendly environment of Hydraulic Engineering Department of UNESCO-IHE. I have learnt a lot from the staff members and my PhD colleagues. I am thankful for their support and help, which they provided in different ways during the course of my study. I express my sincere thanks to UNESCO-IHE for providing me this unique opportunity to work in a multicultural and multiethnic environment.

I am grateful to my employer, Central Water Commission, India, for granting me necessary leave for conducting this study at UNESCO-IHE.

Finally, I am thankful to my parents, my wife Reetu and my son Vibhor for their love and support.

To my family....

Contents

Abstract

Hydraulic pipe networks are an essential hydraulic infrastructure of all the systems involving supply and distribution of water. These networks are used for transmission, distribution and supply of water and, thus, play an important role in many branches of hydraulic engineering. Long-distance pipeline transportation of water for industrial and domestic purposes, pipe networks for city water supply, industrial piping systems, pipelines used for water conveyance for hydropower development etc. are few examples of applications of these networks. They also feature in many irrigation schemes, in waste water disposal schemes and pumping stations.

With the increasing population, urbanisation and industrialisation, the demands on pipeline transportation of water are increasing everyday. Not only more and more new hydraulic networks are to be planned and designed but the existing networks are also to be rehabilitated and expanded. Optimum use of the capacity of the existing networks by efficient operational measures is essential to reduce future investment and to avoid wastage of money through failures.

Growing demand and complexity of hydraulic networks has led to the constant research efforts by engineering community for efficient design and management of these systems. Area of analysis and control of networks has been intensively explored by researchers, however, expected results have not been achieved. Reasons for this may be attributed to the physical and topological complexity of these networks and development of non-coherent approaches to solve different types of network problems associated with steady and unsteady flows. In fact, scientific community has viewed different network problems from different perspectives. Considering the complexity of network configuration and constraints together with degree of complication of objectives to be met, a more unified and coherent approach to the network design and operation is needed.

Problems of network analysis, design and operation can be dealt by two approaches; analysis approach and synthesis approach. Analysis approach uses trial and error procedure to evaluate design variables. Most of the current codes are based on analysis approach. Synthesis approach has not been fully explored, aims at direct evaluation of design variables for the specified hydraulic behaviour of the system. This study provides a general network synthesis framework for approaching network design and control problems from synthesis perspective.

In this study, first a generalized network model has been developed. A network consists of different types of components, constraints and demands. All these represent different kind of boundary conditions, which are mutually interacting. This study has systematically categorized these different boundary conditions into six types of elements. These six types of elements are classified based on known/unknown values of head loss and discharge variables. Study shows that this division of different boundary conditions into six types of elements is comprehensive in the sense that a variety of possible practical problems can be modelled by these six types of elements. In this study, these six types of elements have been given name as element type 1, 2, 3, 4, 5 and 6.

Synthesis approach to network analysis, design and control aims at design of system variables for the specified hydraulic behaviour of the system. For a well-posed problem and its unique solution, these specifications and design variables cannot follow arbitrary topology. The study has developed conditions for a well-posed problem and network solvability in terms of necessary and sufficient conditions for the existence and uniqueness of a solution. The seven necessary conditions and one necessary & sufficient condition, which are defined by eight theorems, must be satisfied by a network model of any problem.

These conditions put restriction on topological allocation of specifications and design variables. Briefly, these conditions can be stated as : one known head, no loop 1 & 3 no loop 1 & 4, no cut 2 & 3, no cut 2 & 4, $ne3 = ne4$, existence of T_{13a} and T_{14a} and non-singular matrix M $(=r_b + L_{ba} r_a L'_{ba}{}^T)$. Study shows that these conditions, which have been developed, using graph theory, are helpful in demarcating different network properties. One of the major results of this study is the development of the necessary condition about existence of two spanning trees, T_{13a} and T_{14a}. Its usefulness has been shown in mathematical formulation and development of algorithm for solution of problems concerning both steady and unsteady flows. Algorithm for generating these spanning trees has been developed in the study, which is easy to implement on computer.

The current study shows that synthesis approach to network analysis and control has a vast potential in the area of system design, transient design, transient system component design, operational optimization etc. In the study, different kinds of network problems concerning steady and unsteady flows have been classified systematically and have been treated in different chapters. Different problems have been viewed from a common platform, which is based on network synthesis framework. Applications of the developed procedures are shown in the area of network analysis, design, control and calibration.

In the study, problems concerning steady flow have been divided into two types: analysis problem and analysis design problem. Using the result of existence of spanning trees, T_{13a} and T_{14a}, the problem formulation is well structured. Mathematical formulation using this condition provides automatic separation of independent and dependent variables. Moreover, a minimum set of independent variables is obtained automatically. The other advantage of the algorithm is that it provides explicit calculation of system parameters, as these parameters are not part of independent variables.

The developed algorithm with the above stated advantages among others is a powerful tool not only for efficient solution of engineering problems, but also for learning and understanding the interrelationships between different variables and hydraulic behaviour of system in general

The non-linear programming formulation of steady state problems is carried out using optimisation approach. Formulation has developed objective functions, which are based on content and co-content of the network. These objective functions are convex in nature and solution of network problem is obtained by minimisation of these functions. Development of this method provides an algorithm for network solution, which guarantees convergence to the sought solution.

Study shows a variety of applications of analysis and analysis-design problems in network design, operation and calibration.

Conventional approach to transient control and transient system component design uses trial and error procedure of analysis approach. The present study has developed procedures for control of transients and system component design using synthesis approach. The procedure can be termed as transient design. Transient design and transient system component design aims at design of boundary conditions and system components for the specified transients in the system. Procedures have been developed for both pressure surges i.e. transients analysed using rigid model and transients analysed using elastic model.

Control of pressure surges using valve operations aims at design of valve operations for the specified transients. Problem can be defined as design of valve operations for transferring a network from an initial steady state to a final steady state within a specified time. At the end of time T, valve operation ceases. Presence of residual transients depends on the network topology. In the study, a criterion for network controllability has been developed. As per this criterion, for the full control of pressure surges, number of valves in the network should be equal to number of loops in the network and their location should be in the chords of the network. Networks following the above criteria can be transferred to another steady state without any residual transients.

Based on number of specifications and design variables along with the topology of the network, the study has divided systematically control problems into four types. These four type of control problems are determined full control problem, determined partial control problem, underdetermined full control problem and underdetermined partial control problem.

Determined problems use only a numerical technique for the solution of a set of ordinary differential equations. However, formulation of these problems is not straightforward. Network model must follow necessary and sufficient conditions for the existence and uniqueness of a solution. The current study provides algorithms for the solution of these problems. Problem formulation utilises network topological properties in the sense that it utilises the presence of two spanning trees T_{13a} and T_{14a}. Algorithm provides automatic separation of independent and dependent variables and provides a minimum set of ordinary differential equations describing the hydraulic behaviour of the system. One of the main advantages of the algorithm is that valve operations are not part of independent variables and these are calculated explicitly.

Underdetermined problems of pressure surges control utilise minimisation of an objective function. In the study, a quadratic objective function is developed which aims at minimisation of change in head and time rate of change of head in the system. Procedure is developed using the principles of calculus of variations, which provides analytical solution of the problem.

Study shows a variety of applications of these developed procedures in the area of transient control, operation optimisation, on-line control and system design.

This study provides a generalised algorithm for transient system component design and shows its application in the design of surge tank and valve operations. Algorithm utilises optimisation techniques and develops procedures for the direct design of these components. Traditional approaches for transient system component design uses trial and error procedures. The present algorithm avoids trial and error procedure and provides design parameters of components directly for the specified transients in the system. These procedures are helpful in system design and optimisation and provides an insight into the relationships between different components in the system.

For the control of transients using elastic model, study highlights the importance of using rigid model results. For cases where change in flow is slow, transient control can be done using the results of rigid model. Valve operations obtained using the rigid model can well be applied in such cases to control transients.

The study has extended the valve stroking principles to include looped networks. In looped networks, for valve stroking to be possible, there should be a valve in every loop and its location should be at the extreme end of each chord. In fact, in most of the practical cases, presence of residual transients is not objectionable. The study has developed a procedure for transient control with the specified residual transients or reduced maximum change in head in a simple system. Allowing residual transients further reduces the head in the system.

The present study has developed the concept of network synthesis and has shown its application in system design, transient design and transient system component design. Study shows that the developed approach is helpful in solving a variety of engineering problems related to network analysis, design, control and calibration.

Notation

a	wave speed
A	cross sectional area of pipe
A_s	area of surge tank
A_v	area of opening of valve
B	a/gA
Cd	coefficient of discharge of valve
D	diameter of pipe
f	Darcy-Wiessbach friction factor of pipe
F_1	$(dH^*_{N\,max}/dt)/2|\varsigma|$
F_2	H^*_N/ς
g	acceleration due to gravity
G	directed graph
h	head loss
H	Head
H_R	Reservoir head
H_s	Head in surge tank
K	$m_b + L_{ba}m_aL'^T_{ba}$
K1	$L'^T_{2a}q^*_2 + L'^T_{3a}q^*_3$
K2	$L_{b1}h^*_1 + L_{b3}h^*_3$
l	length of pipe
L	loop incidence matrix
m	l/gA ; inertia of pipe
ne	no. of elements
nl	no. of loops
nn	no. of nodes
N	node incidence matrix
q	discharge in a element
Q	outflow at node
R	$fl/2gDA^2$; resistance of pipe
t	time
T	total time of operation
T_e	2l/a
T_w	lv_R/gH_R

T_{13a}	spanning tree with elements type 1, 3, a
T_{14a}	spanning tree with elements type 1, 4, a
u	$\gamma/2$
v	velocity
x	distance

α	weight for dh / dt
β	weight for h
γ	eigenvalue
θ	T/T_e
ρ	$av_R / 2gH_R$; Allievi's elastic parameter
Γ	θ/ρ
ε	relative error between elastic and rigid models
τ	dimensionless valve coefficient
ϕ	objective function
ψ	objective function
λ	Langrange multiplier
ς	$(m_R\, q_R\, /TH_R)N_a^{T-1}\, m'_a\, L_{4a}^T\, \Delta q'_4$

Subscirpts

1,2,3,4,5,6,a,b	element types
c	chord
D	dynamic component
N	node
Max	maximum
o	initial value
R	reference
s	surge tank
t	tree
T	time T
v	valve

Supercripts

T	transpose of a matrix
*	known value of variable

1 | Introduction

1.1 Introduction and Motivations

A modern society is to a large extent a system of networks for communication, transportation and the distribution of energy and goods. Networks of pipes, conductors, beams and columns etc. are in constant use in the fields of hydraulic engineering, electronics, electrical engineering, structural engineering and others. The complexity and cost of these networks demand that existing networks be effectively used and the new networks be rationally designed. To meet this demand, there has been a continuing research efforts by the engineering community for the development of procedures and tools for the effective analysis, design and operation of these networks.

Hydraulic pipe networks in engineering applications

Hydraulic engineering has served mankind throughout the ages by providing drinking water. It is amazing how early in civilisation the efficacy of the pipeline was recognised and utilised. The Chinese are believed to have piped water through bamboo lines about 5000 B.C., By about 200 B.C., Rome had a water system that handled 332 million gallons of water per day, in which most of the small diameter pipe was of lead. The first iron underground system of pipelines is believed to be constructed in Paris in 1685.

With the increasing scientific developments and the demands generated by urbanisation and industrialisation, use of pipeline systems for supplying water for industrial use and other purposes has been fully appreciated.

Hydraulic pipe networks are an essential hydraulic infrastructure of all the systems involving supply and distribution of water. These networks are used for transmission, distribution and supply of water and, thus, play an important role in many branches of hydraulic engineering. Long-distance pipeline transportation of water for industrial and domestic purposes, pipe networks for city water supply, industrial piping systems, pipelines used for water conveyance for hydropower development etc. are few examples of applications of these networks. They also feature in many irrigation schemes, in waste water disposal schemes and pumping stations.

Water supply systems for domestic use are the most common example of hydraulic networks. In developing countries, water supply is scare. With the socio-economic development, the work of water supply industries in these countries will increase. There is need for efficient operation of the existing systems and economical expansion for the increased demands. Proper planning and optimal procedures for design and operation of these networks are must to reduce capital investment and operational cost of these networks. Apart from drinking water supply, pipeline systems are associated with industrial water supply, wastewater discharges and the passage of runoff floodwaters through urban areas.

The last decades urbanization and sanitation developments have resulted in larger and more centralized sewer systems, which have increased the need for distant transport of sewerage. Accordingly a large number of pumped sewer mains have been established. This trend is expected to rise further. In contrast to traditional sewer pipeline practice, pumped plastic mains are laid in length profiles directly following terrain. The hydraulic design of these pipelines will also include an analysis of transients and to restrict pressure rise efficient operational procedures.

With the increasing industrialization, demand for energy is increasing specially in developing countries. This has lead to fast development of new power plants and expansion of existing ones. In hydropower plants, water is conveyed through a pressurised system to the machine. The systems are subjected to high variations of pressure change and are, thus, required to be designed and operated efficiently.

Design and operation of pipe networks: research and development needs

Pipeline systems are generally complex both in the physical and topological sense. Capital cost of these systems is generally high. As such, a great deal of attention must be paid to their planning and management. This requires rational design and efficient operation of these systems.

A reliable analysis tool is a prerequisite for the rational design and good operation of piping systems. Computer-aided analysis of the flows and pressures in these networks is now an integral part of the design and management of these systems. Although a number of solution strategies and a large number of computer programs exist that can be used to perform this task, it is quite common to find that for some specified hydraulic behaviour of the system, a model becomes ill determined and fails to provide any solution. Reasons for this have been attributed to the non-convergence and non-uniqueness of the solution.

A systematic analysis tool for hydraulic pipe networks was first given by Hardy Cross in 1936. Hardy Cross method is iterative, slow and suffers from convergence problem. Later on, in 1963 Martin and Peters used Newton-Raphson method for network analysis. By this method, number of iteration steps gets reduced considerably, however, convergence problem remains. This led to research into Newton-Raphson method and few techniques were developed which provided faster solution and better convergence. In 1972, Wood and Charles developed Linear Theory method. In this method also, convergence of solution is not guaranteed. In 1978, Collins used optimization theory for network analysis. Though, the method provides guarantee for the solution convergence, but it lacks generality in the sense that method is limited to networks containing some simple elements such as pipes with known resistances and can not be applied to networks containing other elements such as pipe with unknown resistance, flow or pressure constrained element etc. Much research has been carried out in the last few decades to improve the convergence and solution speed in network analysis tool. Still a methodology, which guarantees the convergence of the solution for a generalized network, is missing.

If a network contains some known and some unknown parameters, their topological distribution in the network should be such that number of independent equations should be equal to number of unknown parameters. Otherwise, network analysis will not provide unique solution. Question of network solvability and solution uniqueness has been addressed by few researchers. Shamir and Howard in 1968, Gofman and Rodeh in 1981 and later on Bhave in 1990 developed some rules without any mathematical proof for network solvability. These rules lack generality of networks and are not complete and even found to be not correct in in some cases. Unfortunately, none researcher has been able to provide necessary and sufficient conditions for the existence and uniqueness of the solution for a generalized network. This may be due to great complexity of these systems arising because of large number of different kinds of components. Determination of these conditions is necessary for the effective and reliable modeling of these networks. Guaranteed unique existence of a steady state for some specified or desired hydraulic behaviour of the system is a prerequisite for the reliable design and operational control.

Design of pipe networks involves methods and procedures to deduce the best configuration of pipe networks and components, for new systems, and also to cater for optimal expansion, of existing systems. In each case the specified performance should be obtained for a minimum investment cost whilst allowing for efficient operation over the expected range of operating conditions. Cost optimal design of networks has received much attention in the last two to three decades and numerous methodologies based on linear programming, non-linear programming and evolutionary techniques (genetic algorithm) are available. However, less attention has been paid to the direct design of system components for the specified hydraulic behaviour. Especially for the expansion of the existing system, it is often required to design new components while meeting some hydraulic requirement of pressures and flows at some locations in the old network. Algorithms providing direct design of such components are required to avoid trial and error procedures.

Pipelines rarely operate in steady state conditions. Anticipating and controlling transient response is a critical design activity for ensuring both the safety and integrity of distribution systems and their effective operation.

The general theory of hydraulic transients was established primarily by the remarkable contributions of Allievi and Joukowsky in the late nineteenth and early twentieth centuries. The same was extended in the twentieth century by Jaeger, Streeter and Wylie among others. Today, transient analysis of pipe systems is very common. However, the field of operational control of transients requires more attention for the safety of these systems.

Operation concentrates on optimal control of piping systems. In essence it requires development of best operating strategies. Mathematically, operation of a system is carried out to transfer it from one steady state to another steady state. Operation of the system causes transients and uncontrolled pressure and flow surges can easily exceed the specified safety limits of the network components. Control of transient effects is important since it can cause potentially hazardous situations. Control of transients through rational valve operations is considered to be the best solution in all sectors related to water, gas, oil and other fluids.

Development of optimal control methods involves evolution of operational strategies for the specified transients. This is called 'transient design' and the methodology is termed as 'design of operation measures'. Developments in this area were carried out in 1960s and 1970s mainly by Streeter, Propson and Stoner. Technical developments in this area are limited to some very simple systems. Reason for this is attributed to the complexity of network configuration and transient phenomena itself. For large networks, still iterative techniques are used to obtain the best operation strategy. Development of methodologies for the optimal control of transients in large networks is necessary for the safety of the components, proper utilisation of network component capacities and cost effective design.

Design of system components that are used for controlling transients is carried out using trial and error procedures. One typical example is the design of a surge tank. A surge tank and other network components are designed to meet specified hydraulic behaviour of the system. An immediate question that arises in the mind is could there be an approach that provides design parameters of the components directly for some specified hydraulic behaviour of the system. Development of such approach would be helpful in avoiding trial and error procedures and would provide efficient analysis tool for the problems related to design, operation and performance.

Network synthesis approach to design and operation of pipe systems

Although the increasing availability and the decreasing cost of digital hardware and software has lead to an effort aimed at the development of sophisticated software for the analysis, design and operation of pipe networks, unfortunately in most cases those efforts have not produced the expected results. Scientific developments show that researchers have approached network problems from different perspectives resulting in the development of various methodologies that are not coherent in dealing with different kinds of problems. The main reason for this is often attributed to the excessive physical and topological complexity of these networks leading to non-generalised models, often encountering of ill-determinacy and iterative trial and error procedures.

Pipe networks may be divided into three parts: components, boundaries and constraints. Reservoirs, valves, surge tanks, pumps, pipes etc are some common components of a network. Valve settings, pump operating points, reservoir levels, demand etc defines boundaries of a network system. Constraints are generally physical and hydraulic. Maximum/minimum valve openings, pump capacity etc. are physical constraints. Hydraulic constraints are related to limitations on head and flow at different locations of a network. Network design and operation problems involve component design and design of operation measures to meet the demand while keeping the constraints satisfied. Traditionally, this task is solved through analysis approach that involves trial and error procedure. To avoid these cumbersome procedures, a network synthesis approach which designs network components or operation measures for the specified or desired hydraulic behaviour of the system directly is essential.

Network synthesis implies design of components and boundaries of a system with the given sets of inputs to meet the specified outputs by keeping the system constraints satisfied. It incorporates both component design and design of operation measures for the specified system behaviour.

Motivations

With the increasing population, urbanisation and industrialisation, the demands on pipeline transportation of water are increasing everyday. Not only more and more new hydraulic networks are to be planned and designed but the existing networks are also to be rehabilitated and expanded. Optimum use of the capacity of the existing networks by efficient operational measures is essential to reduce future investment and to avoid wastage of money through failures.

An overall goal for piping systems is to distribute source supplies to consumers as safely, securely and as economically as possible. This requires a rational design and good operation of these systems. Capital costs for these networks are usually very high and it is therefore sensible that they are designed and operated to give maximum value for the money invested.

Pressure pipe systems are subjected to a wide range of operational requirements. If operational measures are not properly planned, it can cause high pressure rise resulting in the failure of the system. In Canada alone, the estimated cost of repairing water main breaks exceeds Can$100 million annually (Karney and McInnis, 1990). Development of tools to control these pressure rises in the systems is essential for avoiding failure and cost. Hence, it is necessary to concentrate research efforts towards the development and management of pipeline system that has dependability, safety, acceptance and efficiency.

Considering the complexity of network configuration and constraints together with degree of complication of objectives to be met, a more unified and coherent approach to the network design and operation is needed. That was the main premise for starting this research.

Design of pipie network components and design of operational procedures to reduce the effects of transients has received much attention. Generally, the developed procedures are iterative. Synthesis approach is required for network design and operational control problems. Network synthesis approach implies design of system components and operational procedures of the system directly for the specified system behaviour by keeping the constraints satisfied. This is the general, network synthesis, context in which this research has been conceived.

1.2   Network Flow Models and Network Problems

Pressurized pipe systems are subjected to a wide range of operational and loading conditions that vary with time. Analysis of these conditions is required for the design, as well as a safe and efficient operation.

The network flow models are typically classified (Fig. 1.1) as dynamic models, which consider flow rate and pressure as time-dependent, and static models, which consider flow rate and pressure as time-independent. Dynamic models can be further classified as inertial models that take into account system inertia and non-inertial models in which the dynamic characteristics depend on the variability of the boundary. The two types of inertial models are considered as water hammer or elastic models (distributed system approach) and as pressure surges or rigid models (lumped system approach). Non-inertial models are typically referred to as quasi-static models or extended period simulation models.

In short, four different models are available for analysing flows in piping systems: static model, quasi-static model, elastic model and rigid model. Application of these models depends on the type of problem.

Static model is used for steady state simulation. Model computes flow rates and head losses in all network components for the given network characteristics and demands. If a network component is to be designed to produce a specified head and flow rate at a given location, trial and error procedure is used.

Quasi-static model is used to simulate hydraulic behaviour of the network for the given network characteristics and demands which change time to time. In fact, this model carries out successive steady state simulations taking into account the available storage and the change in demands. Model is used when demand changes on hourly or daily basis and, thus, inertial effects can be neglected.

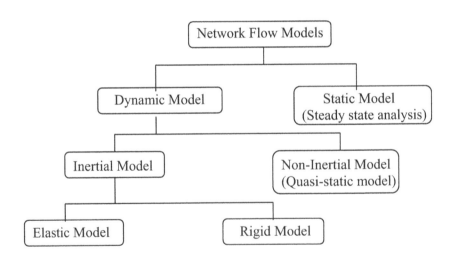

Fig. 1.1 Network flow models

Elastic model is used to simulate hydraulic behaviour of the system subjected to fast change in boundary conditions. Model is based on distributed system approach and analyses water hammer effects caused by change in boundary conditions.

Rigid model is based on lumped system approach and analyses pressure surges in the system. Rigid model is generally used when change in boundary conditions is slow. Though rigid model is an approximate model and exact transient effects are simulated by elastic model, it is frequently used for the cases where results of two models show little error or error within acceptable limits.

In general, different network problems can be grouped in two categories namely planning and design and operation and control. These two problems are interrelated in the sense that planning and design of networks is carried out based on operational range and operation procedures and operation of networks is done based on the design criteria (Fig. 1.2).

Planning and design problems involve selection of best layout and safe design of the network components. Safe design looks at some of the factors, which must be taken into account to result in systems, which are intrinsically safe to operate. Consideration is largely restricted to the effects of fluid transients. Uncontrolled pressure and flow surges can easily exceed the specified safety limits of the network components. Such transient effects are important since they can be inadvertently initiated by operator actions, or component failures, and can cause potentially hazardous situations.

In general, every network problem involves static flow model and one dynamic model. For example, if in a water supply system, some slow changes in valve settings are required for daily demand variation, the design and operation problems will involve static model and a dynamic model such as non-inertial model or rigid model.

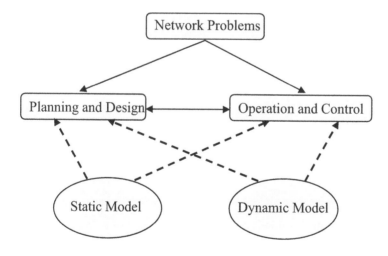

Fig. 1.2 Network problems

Though problems related to planning and design of a network are inter-linked with operation and control problems, the subject area in literature has been dealt at two platforms: problems related to steady state flow and problems concerning unsteady flow.

The researchers have generally considered two types of network problems related to steady flow: standard analysis problem and cost optimal design problem. This has lead to development of various analysis codes and cost optimal design algorithms based on different numerical techniques. Development of procedures for direct design of system parameters for a specified hydraulic behaviour of the system has not been given much attention.

Network problems related to steady flow can be approached by two ways (Fig. 1.3). First is the analysis approach in which trial and error methods are used to obtain design parameters of a component for some specific requirement. The second is the synthesis approach, which aims at designing system components directly for the specified requirements.

As shown in Fig. 1.3, analysis approach provides standard analysis of a network and iterative design of its components. Synthesis approach provides direct design of system components for some specified hydraulic behaviour.

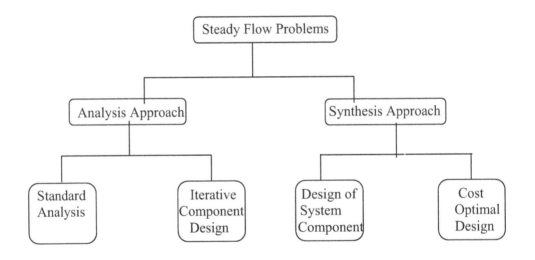

Fig. 1.3 Operation and control problems

Operation and control problems involve transferring the network from an initial steady state to a final steady state. The main objectives during operation may be minimisation of pressure rise, minimisation of pumping cost or any other specified criteria. In general, the problem is to determine the best operation strategy to satisfy the current operating requirements.

Similar to steady flow problems, there are two ways to approach operation and control problems: analysis approach and synthesis approach (Fig. 1.4). Analysis approach involves trial and error methods for the design of valve operations or a system component. However, synthesis approach involves design of valve operations and system components for the specified transients or flow variations directly.

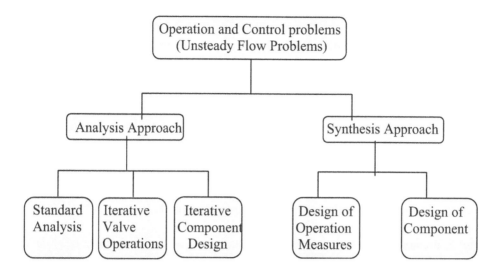

Fig. 1.4 Operation and control problems

1.3   Scope of the Study

The subject area of analysis and control of flows in networks is too vast to be dealt in detail in one study. The scope of the present study is limited to some aspects of steady and unsteady flows in networks.

The study focuses on development of synthesis approach for the analysis and control of networks and its application to the following problems:

• Steady state analysis and system component design

• Pressure surge analysis using rigid water column theory and transient system component design

• Pressure surges control using rigid water column theory

• Transients control using elastic water column theory

Different types of network problems have been considered to show the applications of the developed approach. Cost optimal design is not considered within the scope of present study.

The study concentrates on the development of procedures and methodologies for different network problems. Use of existing numerical technique has been done wherever necessary. Development of new numerical technique is not within the scope of present study. Detailed objectives of the study are presented in the next chapter.

1.4   Structure of the Thesis and Conventions Used

Chapter 2 describes latest scientific developments in the area of analysis and control of flows that falls within the scope of present study. Detailed objectives of the current study are then given and a framework of network synthesis approach is developed.

In chapter 3, first, a categorisation of different kinds of boundary conditions is made. A network model is developed using principles of graph theory and related matrices and vectors are defined.

In chapter 4, first, a critical review of existing network solvability rules is made and then necessary and sufficient conditions for the existence and uniqueness of a solution in a arbitrary network are developed employing concepts of graph theory.

In chapter 5, methodologies for network steady state analysis problems have been developed using principles of optimization and applications are shown by selecting few examples.

In chapter 6, first principles of control of pressure surges are discussed and then a formulation for pressure surges analysis in a arbitrary network is presented. Concepts of network controllability are developed. Applications of the developed methodology for the control of pressure surges are shown by selecting examples.

In chapter 7, methodologies for optimal control of pressure surges by valve operations are developed using optimization techniques and applications are shown through few examples.

Chapter 8 presents the concepts of transient system component design. Methodologies for direct estimation of design parameters of surge tanks and valve operations are presented.

In chapter 9, some aspects of control of transients analysed by elastic water column theory are presented.

In chapter 10, general conclusions of the research as well as recommendations for further research are presented.

At the end of chapter 10, appendices are given which contains reference material to the chapters in which they have been referred.

The relevant variables used in this thesis are summarized in the attached list of notation. Multiple use of the same variable name with a different meaning is avoided as much as possible. Some equations have a very specific naming convention though. In those cases it happens occasionally that a variable is used for the second time with another meaning. However, this is explicitly indicated in the text and these variables are not incorporated in the list of notations.

2 | **Background and Objectives of the Study**

In this chapter, a detailed review of latest developments in the area of analysis and control of flows in networks is done after intensive literature review and some research issues are highlighted. The review of latest scientific developments is carried out for the problems that fall within the scope of the present study. Next, a detailed outline of the objectives of the current study is presented.

At the end of the chapter, a framework of network synthesis approach is developed which is the foundation of present study.

2.1   Analysis and Control of Flows: *Developments and Issues*

In this section, a detailed review of the latest developments in the area of analysis and control of flows, which fall within the scope of the study, is carried out by dividing the area in two categories: network problems associated with steady flow and problems concerning unsteady flow.

2.1.1   Hydraulic simulation of steady state and system component design

Pipe network analysis, wherein the steady state flows and the hydraulic heads are determined in some arbitrary network, is a fundamental engineering tool widely used in all the sectors related to supply and distribution of water, oil, gas and other fluids.

Two types of network analysis problems exist. First is the "analysis or forward problem". In this problem flows and pressures in each system component are determined for known component characteristics and demands. Analysis of an existing distribution network is an example of such an analysis problem.

Second is the "analysis-design or inverse problem". In this problem, some of the system characteristics are unknown and are to be determined to meet the specified flows and/or pressures in some of the components. Expansion of a network is an example of such type of problem. This problem seeks evaluation of elements' parameter such as diameter, roughness etc. for some specified hydraulic requirements. Designating the elements' characteristics as unknowns allows the analysis algorithm to solve directly for these values. However, modelling of such problem is not straightforward. Problem solvability depends upon the topological distribution of known and unknown parameters.

Although a number of solution strategies and number of computer programs exist for solving network analysis problems, questions of solution uniqueness and avenues for improvement in algorithms for better solution speed, convergence and explicit estimation of parameter, still remain.

Uniqueness of solution

The unknown parameters in a network can be element head loss, element flow and element parameter such as diameter, roughness etc. For the network analysis to be feasible, the number of unknown parameters and their distribution over the entire network must be such that an adequate number of independent equations can be formulated. If the unknown parameters are concentrated in one part of the network and the known parameters in the other, it is not possible to determine all the unknown parameters. Therefore, the number and mutual distribution of the unknown and known parameters should obey certain rules so that the network analysis is meaningful.

Few researchers have considered the problem of solvability of networks. Shamir and Howard (1968, 1970, 1977), Gofman and Rodeh (1981) and Bhave (1990) have developed some rules for the network solvability. But these rules have been developed heuristically without any mathematical proof for their validity. It is found that these rules are not complete and loose generality of networks. In some cases, rules are found to be incorrect. A critical review of these rules is carried out in chapter 4 wherein solvability rules for a arbitrary network in terms of necessary and sufficient conditions for the existence and uniqueness of a solution are developed.

Recently, Altman and Boulos (1995) presented the sufficient condition for the solvability of flow constrained network models. According to them, only chords of a spanning tree can be flow constrained. Inclusion of other types of elements such as pressure constrained element, pressure & flow constrained element was not considered.

Collins et al. (1978) and Todini and Pilati (1988) have used optimisation principles to prove that a unique solution to the hydraulic analysis of a pipe network exists if all the network components have strictly convex content function. Solvability rules for an arbitrary network were not presented.

In 1980, Collins, through an example network containing two PRVs (pressure reducing valve) concluded that networks containing feedback devices such as pressure reducing valves, pressure sustaining valve etc. have multiple solutions. In the example two PRVs are both set to control the pressure at the same downstream node. Not other types of components were considered except pipes and thus the results can not be generalised. Chandrashekhar (1980) acknowledges the possibility of non-unique solution. Gessler (1981) suggests that networks with PRVs will have unique solution, but dose not offers any proof. Salgado et al. (1988) assert that "because pressure controlling devices are in essence variable resistance devices, their steady state solution must be unique". The argument given by Salgado is contradictory to the Collins example network showing multiple solution. More recently, Berghout and Kuczera (1997) concluded through examples that networks containing feedback devices will have unique solution provided these devices are connected to pipes in the downstream.

In all the studies related with feedback devices, researchers failed to consider a generalised network model. Moreover, no mathematical proof was given by any researcher in support of presented arguments.

Development of necessary and sufficient conditions for the existence and uniqueness of a solution for an arbitrary network consisting of all possible types of components is still a main problem to be solved.

Algorithms and convergence of solution

Many methods exist for hydraulic network analysis and a number of significant improvements have been made in recent years for solving flows in networks based upon known methods from numerical analysis and graph theory and more are bound to come. Recent mathematical methods coupled with sparse matrix techniques have greatly facilitated the steady state flow analysis of large networks. It has been observed, however, that the iterative procedure for the solution dose not always converge or may sometimes converges to a poor solution.

Analysis of hydraulic pipe networks involves solution of non-linear equations governing the flow. Hardy cross was the first to suggest a systematic iterative procedure for network analysis in 1936. The approach was originally based on loop flow correction equations, i.e. Δq equations known as the method of balancing heads. Later, Hardy Cross described a second method which was based on nodal head correction equations, i.e. ΔH equations. This approach is also known as method of balancing flows. Hardy Cross method is simple and easy to apply but its greatest drawback is its convergence problem. Because the adjustments are computed independent from each other, convergence problems were frequently noted. Convergence is very slow and large numbers of iterations are required while using this method especially for large networks. In some cases, solution starts to diverge or dose not converge well.

Martin and Peters (1963) were the first who used Newton-Raphson technique for the network analysis using nodal formulation. Newton-Raphson method expands all non-linear terms through partial differentiation in Taylor's series and considers the linear terms only for solution by neglecting all higher order terms. This method attempts to solve all the correction equations simultaneously including the effect of adjacent loops which usually reduces the number of iteration steps considerably. Although the authors advocated the procedure, convergence problems were reported when the flow in a pipe was close to zero. The node method was also formulated by Collins and Johnson (1975). The authors acknowledged that 'if a large pipe of short length and relatively low flow exists, many iterations are necessary' before an acceptable solution can be achieved. Convergence difficulties were also reported by

Collins and Kennington (1977), Wood and Rayes (1981), Nielsen (1989), which they linked to pipes with relatively low resistances.

To improve the rate of convergence in Newton-Raphson technique Lam and Wolla (1972) suggested a new algorithm that avoids evaluation and inversion of the Jacobian in each iteration. The method, called modified Newton-Raphson technique is based on the residue of the functions and uses an iterative equation to update an approximation to the Jacobian and its inverse in each iteration. This approximation to the Jacobian inverse is then used to correct the variables so as to produce a smaller residue for the set of equations. Though this method improves convergence, but it is not guaranteed.

Both Hardy Cross and Newton-Raphson techniques require an initial balanced set of flow rates and the convergence depends to a degree on how close this initial set of flow rates is to the correct solution. If the initial estimates are poor, a large number of iteration steps are required to converge and in some cases, the process may still diverge.

Besides convergence, some computational problems are also visible in the algorithm based on Newton-Raphson technique especially in analysing large networks. The method requires inversion of a matrix, called the Jacobian, which turns out to be the most expensive part of the algorithm. If nodal formulation is used, the matrix is of the order equal to the number of nodes, nn, in the network and inversion of the matrix requires nn^3 mathematical operations on the computer and nn^2 memory locations to store it. To reduce the memory size, Epp and Fowler (1970) suggested a loop numbering technique that produces a banded symmetric matrix. Chandrashekhar and Stewart (1975) also suggested a sparsely oriented lower and upper triangular decomposition technique for reducing the memory size while handing large networks.

In 1972, Wood and Charles presented Linear Theory method. This method solves the entire set of hydraulic equations simultaneously after linearising the non-linear terms by merging a part of it in pipe resistance. Method has also been used by Collins and Johnson (1975) and Issacs and Mills (1980). Though this method is considered most stable, convergence of solution is not guaranteed. Moreover, this method takes more number of iterations as compared to Newton-Raphson method. Nielsen (1989) suggested the use of Newton-Raphson method combined with Linear Theory method as a robust starting procedure. This reduces the number of iterations but convergence is not guaranteed.

Collins et al. (1978) used optimization theory for network analysis where the solution corresponds to the minimum of a highly structured and constrained non-linear convex optimization problem. The work showed that the laws of conservation of energy and nodal mass balance were satisfied at the minimum value of a convex non-linear optimization problem. Collins defined content function of network elements and the problem seeks minimisation of a total content of the network. Linear programming was used for the solution. Convex non-linear programming provides guarantee for the solution convergence. However, Collins's work was limited to networks containing some simple elements only such as pipes with known resistances, demands and reservoirs. Other elements such as element with unknown resistance, flow and pressure constrained element etc. were not taken into account.

The reliability of the algorithm employed for pipe network analysis is of great importance. Failure to obtain a solution is an inconvenience and the failure to recognise a poor solution may be even a greater problem because this may lead to poor design or management of network systems.

Today, a number of computer programs are available for carrying out steady state analysis of pipe networks, which use different methods for obtaining the solution. With the increasing hardware capabilities, steady state analysis of large size networks is no more a great problem considering memory requirements and simulation time. However, convergence problem is persistent. This has created much inconvenience to software users who most of the times have to resort to trial and error procedures for initialising the simulation. Moreover, reliability of results obtained from such simulation becomes doubtful.

There is a need for a reliable algorithm for network analysis, which can guarantee convergence of the solution. Development of such algorithm is necessary for effective and reliable network modelling.

Explicit determination of design parameters in analysis-design problem

Analysis-design problem aims at determination of design parameters of some network components for the specified flows and pressures in some other components. This problem finds variety of applications in design, operation and model calibration problems and has a potential as a powerful design tool.

As the number of known and unknown parameters has to be equal, for each specification of head or flow one design parameter can be designated as unknown. Network analysis directly provides the value of design parameter for the specified heads and/or flows. However, as described earlier modelling of such problem in a comprehensive and meaningful manner is neither easy nor straightforward. Specifications and unknown parameters can not assigned arbitrarily in a network. Existence and uniqueness of solution depends on the manner in which specifications and design parameters are topologically distributed.

Though the idea of solving the hydraulic network equations in terms of variables other than pipe flows or nodal heads is not new, little research has been carried out for the development of this approach. The approach was first proposed by Shamir and Howard (1968). They generalised the node equation formulation to deal with different combinations of heads, demands and pipe resistances. Similar approach was also proposed by Gofman and Rodeh (1981) and later by Ormsbee and Wood (1986). Solution for design parameters is obtained by the addition of one continuity or energy equation to the system of equation for each specification of pipe flow or nodal head respectively.

It is observed that these proposed models lack generality and often become ill determined. Moreover, algorithms are not efficient in handling design parameters. This is because the existence solvability rules are not complete and comprehensive in dealing with such problems. Development of necessary and sufficient conditions for solution uniqueness is a prerequisite for meaningful modelling of these problems. Such conditions provide relationships between design parameters and specifications, which can be used for the development of effective algorithm.

Direct estimation of design parameters has great potential in handling various types of problems related to design, operation and performance of networks. Despite the potential of such an approach, not much development has taken place in this direction. Development of these algorithms can be used in the overall design of network systems.

2.1.2   Operation and control of flows and transient system component design

Operation of networks is generally carried out by valves, which aims at transferring the system from one steady state to another steady state. Two main fields of problems exist related to operation of networks. First deals with transients. Second is associated with the optimal scheduling problems, which aims at determination of cost optimal pumping schedules or valve settings for the daily or weekly demand variations. This problem does not take into account transients and, in fact, is based on steady state models.

Optimal scheduling problems are not within the scope of present study. Scope of the study is limited to analysis and control of transients, both slow and rapid. Developments in this area are presented in this section for both slow and rapid transients.

Analysis and control of transients

A change from steady state flow in a piping system occurs because of a change in boundary conditions. There are many kinds of boundary conditions that may introduce transients such as changes in valve settings, accidental or planned, starting or stopping of pumps, changes in power demands in turbines etc. Transient problems in engineering practice are of significant importance because it can cause excessive pressures, vibration, cavitation and noise far beyond that indicated by steady flow analysis. Excessive pressure fluctuations caused by acceleration or deceleration of a fluid mass are the main concern in liquid system transient analysis.

With the increase of complexity and performance requirements of modern engineering network systems, the control of objectionable transients in the operation is becoming increasingly important in a number of applications such as the design of hydroelectric generating stations, water supply systems, cyclic components of otherwise continuous flow processes and hydraulic systems developed for industrial and commercial applications. Therefore, various devices and/or control procedures such as surge tanks, air chambers, pressure reducing valves, valve stroking etc. are used to reduce or eliminate undesirable transients.

An optimal flow control is desired in hydraulic network systems due to operation and design considerations. Optimal control aims at defining a mode of operation of various appurtenances and control devices so that a desired response is obtained. A desired system response may be, for instance, to keep the maximum and minimum transient pressures within specified limits, changing the flow from one steady state to another without flow oscillations and so on. For example, a valve at the downstream end of a pipeline may be closed in such a way that the pressure dose not exceed a specified limit and the transients are completely eliminated at the end of valve operation. Such a valve operation has been referred to as optimal valve closure or valve stroking.

Optimal flow control is a synthesis approach in which the variations of boundary conditions are determined to obtain a desired system response. This is also called transient design. This approach is different from the usual analysis approach in which the boundary conditions are specified and resulting response of the system is analysed.

The subject area of design of valve operations to control transients and more specifically transient design has not received much attention from the researchers. Scientific developments in this area are limited to control of transients in some very simple systems. Subject area can be divided into two categories; control of pressure surges or slow transients and control of rapid transients.

Analysis and control of pressure surges using rigid model

Pressure surges or slow transients with low frequencies are analysed using rigid water column model (RWCM) which treats the fluids as an inelastic substance wherein pressure changes propagate instantaneously throughout the system and elastic properties of the pipe walls are of no consequence. The rigid water column model is a lumped approximation to the elastic column model (Watters, 1984).

A wide range of pipeline problems falls within the domain of rigid water column theory. Transients generated in the systems consisting of surge tanks and subjected to slow valve operations are of low frequencies and can well be analysed by RWCM. The equations describing this type of flow are generally ordinary differential equations, which can be solved in closed form or with relatively straightforward numerical techniques.

Until now, RWCM has not been applied for pressure surges analysis of complicated piping systems (Watters, 1984; Streeter and Wylie, 1993). It is observed that it is difficult to derive a system of non-linear ordinary differential equations of the first order, which can be numerically integrated.

Though, approaches based on both loop and incidence method are developed for analysing pressure surges in looped networks, authors considered some very simple systems and their algorithms lack generality (Onizuka, 1986 and Shimada, 1989). It is observed that it is difficult to generate generalised model due to complexity of networks. This generally results in a set of ordinary differential equations in which it is difficult to separate independent and dependent variables..

A generalised and efficient algorithm, which can handle arbitrary topology of network and provides a minimum set of ODEs that can be numerically integrated easily, is necessary and needed.

Control of pressure surges by valve operations has not received much attention by the researchers. Shimada (1992) considered the pressure surges control problem and used procedures of optimal control theory. The objective function used was directed to achieve just final steady state valve settings. Neither control of pressures due to transients was considered as an objective nor the control of residual transients at the end of valve operations was given attention. In fact, procedure developed can not be termed as control of slow transients because with the known final steady state valve settings, any number of trajectories of valve operations can be found which will take the system to final steady state.

Control of pressure surges in pipe networks is still a novice area. Design of pressure surges or transient design has not been considered so far. Transient design envisages design of valve operation rules for the specified transients. Development of methodologies for transient design is necessary for effective control of pressures and flows and good operation of network systems.

Analysis and control of transients using elastic model

To analyse rapid transients elastic water column theory wherein the elasticity of both fluid and the pipe walls is taken into account is well developed (Streeter and Wylie, 1993; Watters, 1984; Choudhary, 1979). However, not much research has been carried out for the control of rapid transients by valve operations. Developments in this area are restricted to the works of Streeter (1963, 1967), Propson(1970) and Stoner (1968).

The first study, which succeeded in eliminating the objectionable residual transients, was that of Streeter. Considering a simple frictionless pipe and aided by the visual insight afforded by utilisation of the dependent variable graph (Allievi's graphical method), he presented a procedure by means of which the downstream valve motion can be determined which would create a controlled transient between any desired initial and final steady-uniform flow conditions. More specifically, the flow change is accomplished in three phases: the length of the first and last phase is one round trip wave travel time and the central phase is of such duration that the change can be effected within the specified pressure limits. The head dose not decreases below initial steady state values and dose not exceeds the predetermined maximum. He restricted the valve closure to duration of twice the round trip wave travel time.

Streeter further extended the scope of valve stroking principles by taking into account frictional effects. The method was then readily extended to branching systems by apportioning the flow changes among the branches as a linear relation of the initial to the final steady-uniform values in each pipe.

Later, Propson developed procedure for valve stroking in a specified time and a more rapid closure, down to one round trip wave travel time. These procedures provide optimal control of flows, however, valve operation rules are not smooth and a sudden closure or opening is obtained at the beginning and end of operations. Procedures of valve stroking have been applied to some simple branched systems. Though these procedures are able to provide optimal flow control, it is noticed that valves have to act against a very high head.

Generally, transients control problem involve following aspects:

- Limit on valve operation time

- Limit on maximum or minimum pressure in the system

- Elimination or existence of residual transients after valves stops to operate

Valve stroking is a procedure, which eliminates residual transients, and maximum or minimum pressure in the system is governed by the valve operation time or vice versa. In many cases of network operation, presence of some residual transients dose not effects the requirements of the users. Generally, limit on maximum or minimum pressure is the main concern. If residual transients are permitted, with the same valve operation time limits on maximum or minimum pressures can be lowered. Such procedures of transients control are yet to be developed.

Transient system component design

Design of various system components used for controlling transients is carried out using trial and error procedures. Some typical examples are design of surge tank, air vessel, conveying elements etc.

Conventional procedures for the design of surge tanks consist of assuming a surge tank design and analysing the result of transients due to different planned and accidental cases. If the resulting behaviour of the system is not satisfactory, a new surge tank design is adopted. Procedure is repeated until established criteria are satisfactorily met.

Similar procedures are used for other the design of other components also such as air vessel, conveyance element, valves etc. These trial and error procedures are very cumbersome specially if the network topology is complex.

Generally, these components are designed to meet some specified transient response in some other components. Methodologies providing direct design parameters would be effective in avoiding cumbersome and tedious trial and error procedures. Moreover, such procedures would help in providing an insight into the relationships between the mutual behaviour of different components.

In general, transient design and transient system component design are still unexplored fields. It aims at design of boundary conditions for the specified transients. Developments in these areas can simplify the existing procedures of design and operations.

2.2   Objectives of the Study

The concept of network synthesis is the foundation of the present study. Within the framework of network synthesis, the study aims at development of methodologies for analysis and control of flows in pipe networks. Within this context, the main objectives of the present study are:

1. Development of tools for steady flow analysis and component design using synthesis approach

Scope of the study in this area includes

i)   Development of a generalised network model after categorising different possible boundary conditions which can represent an arbitrary network topology, demands and constraints.

ii)   Development of necessary and sufficient conditions for the existence and uniqueness of a solution to network analysis problems.

iii)   Developments of methodologies for the solution of network steady state analysis and component design problems using optimization methods.

iv)   To show applications of the developed approach by examples.

2. Development of tools for the control of transients using synthesis approach.

Scope of the study in this area includes

i)   To develop the principles of transients control using synthesis approach and to develop criteria for network controllability.

ii)   To develop methodologies using optimization principles for the control of pressure surges (slow transients) by valve operations in networks using synthesis approach.

iii)   To develop the methodology for transient system components design and show applications for the design of surge tanks and valve operations.

iv)   To show usefulness of the developed methodologies for the control of pressure surges in controlling rapid transients in the network by valve operations.

The full accomplishment of these objectives implies the solution of various sub-tasks along the way leading to the development of reliable and efficient tools for network steady state simulation, control of pressure surges, rapid transients and transient system component design.

At all levels potential of network synthesis approach has been shown in viewing different types of network problems encountered in practice from a coherent perspective and in providing their solutions.

2.3   The Network Synthesis Framework

The framework of network synthesis approach has been derived by viewing pipe networks from system-oriented-approach. To elaborate this approach, it is necessary, first, to define the entities related to a network system. From system-oriented-approach, a pipe network can be described in terms of components, constraints and objectives.

Components consist of pipes, hydraulic devices such as pumps, valves, surge tanks etc., reservoirs and consumer demands. Every component has a unique characteristic, which is described by a relationship between head loss and flow through it.

Constraints include the restrictions imposed on the system because of the physical construction of the components and hydraulic performance requirement. Hydraulic performance requires restrictions on pressures and flows.

Objectives encompass the distribution of source supplies to consumers as safely, securely and as economically as possible. Requirement of safety implies restriction on maximum or minimum pressures and velocities or simply minimisation of pressure and velocity changes. Security means an assurance to meet the demands. Objective of economy is to obtain a cost optimum design or operation. This implies optimisation to ensure the best compromise between these conflicting objectives.

An arbitrary network is composed of large number of mutually interacting components. Mathematically, all components and constraints are different kinds of boundary conditions that are bonded together and are mutually interacting. These boundary conditions may be static or time-invariant as in the case of steady state or these may be dynamic or time varying as in the case of unsteady state.

To understand the hydraulic behaviour of a network, it is necessary first to understand the relationships between physical and topological properties of these mutually interacting boundary conditions.

In a network system, some boundary conditions are known and some are unknown. Known boundary conditions are related to the known inputs in the system and the desired hydraulic behaviour. These known boundary conditions are called decision parameters or decision variables or simply specifications. Unknown boundary conditions are those, which are to be designed and are called design parameters or design variables. These unknown boundary conditions are to be designed to meet the specifications. Design of unknown boundary conditions to meet the known boundary conditions is called network synthesis.

A large and complex pipe network consists of arbitrarily distributed different kinds of known and unknown boundary conditions. Mathematical modelling of such systems in a comprehensive and physically meaningful way is neither apparent nor straightforward. The problem of solvability of these models is dependent on the manner in which design parameters and corresponding boundary specifications are distributed over the network. In other words, design parameters and decision parameters can not be put arbitrarily in a network. Distribution of these parameters must follow certain rules to guarantee a unique solution. Determination of such rules is necessary and a prerequisite for the effective modelling of these systems. Otherwise, mathematical models become ill determined. Obviously, solvability rules are dependent upon the relationships between physical and topological properties of different kinds of boundary conditions in a network.

The first step of network synthesis approach is to develop network model for the known network data, specifications and unknown boundary conditions (see Fig. 2.1). This requires categorisation of different boundary conditions.

Second step is to check problem solvability that demands development of network solvability rules.

Third step is the mathematical formulation of a network problem and development of efficient methodologies required for the solution of a problem. Methodology may use only analysis or optimization tools or a combination of both depending upon the problem objectives. Methodology provides desired design parameters.

This study is an attempt to develop network synthesis approach for the solution of various types of network problems. Not all kinds of problems related to network analysis, design, and operation have been considered within the scope of this study. Much attention is given to steady state analysis, control of pressure surges and transient system component design. However, application of the developed network synthesis approach has been shown in viewing the different problems from the same perspective.

This study shows that network synthesis approach finds applications not only in other water resource systems but also in other branches of engineering.

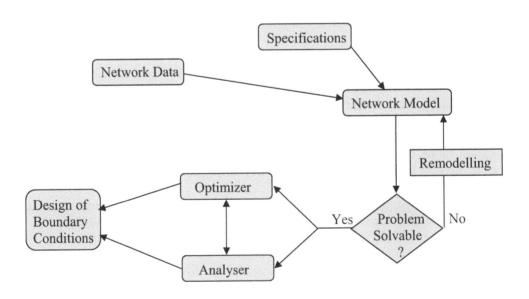

Fig. 2.1 Framework of network synthesis

2.4   Concluding Remarks

Scientific developments in the area of analysis and control of flows exhibit use of analysis approach for the solution of various types of network problems. Development of synthesis approach has not been given much attention.

Synthesis approach aims at designing the system components and boundary conditions for the specified hydraulic behaviour of the system.

In the field of steady flow, literature shows the attempts made by the researchers to develop synthesis approach but expected results have not been achieved. This is due to following reasons:

• Network models considered were not generalised.

• Not all-possible boundary conditions were considered.

• Necessary and sufficient conditions for the existence and uniqueness of a solution were not developed.

Due to these reasons researchers are facing the problem of non-unique solution or sometimes even no solution. Development of procedures for analysing and designing a arbitrary system for the specified hydraulic behaviour needs first categorisation of all possible boundary conditions and development of necessary and sufficient conditions for the existence of a unique solution.

The area, which has received much attention, is the method of solution of system of equations describing a steady flow in the network. Many algorithms have been proposed by the scientific community, however, still a method, which provides guaranteed convergence to a solution is missing.

The field of operation and control of networks has also not received the needed attention by the researchers. Control of transients is carried out by providing components like surge tank, air vessel etc. and by proper valve operations. Current procedures for the design of these components and valve operations are based on trial and error procedures. Though concepts of valve stroking have been developed but these procedures have been used to control transients in some simple systems. Moreover, it is felt that these procedures require valves to act under very high heads.

Similarly, design of system components such as surge tank, for the control of transients, is carried out using analysis approach. Procedures giving direct design of these components for the specified transients are still to be explored.

Though the need to control the objectionable transients is constantly felt but scientific developments exhibit the use of analysis approach or trial and error procedures to solve such problems. Development of procedures for the operation of a system for the specified transients is needed. This procedure is termed as transient design and its development is needed for the effective control and better understanding of the hydraulic behaviour of the system.

In general, synthesis approach for the design of system components and boundary conditions for specified hydraulic behaviour, whether steady or unsteady, has not been fully explored. The approach has good potential for the design and control of networks. The present study deals with the subject area of analysis and control of networks within the framework of network synthesis approach.

3 | A Generalized Network Model

A hydraulic network is a complex system both physically and topologically. It consists of several types of components such as pipes, pumps, valves, reservoirs, surge tanks etc. Each component is described by its constitutive relationship, which relates head loss and discharge through it. Apart from different components, a network problem may be associated with different hydraulic constraints in terms of specifications of head/head loss and/or discharge at different locations. In fact, both components and constraints are different types of boundary conditions, which are mutually interconnected. A hydraulic network may be described as a system composed of different mutually interconnected and interrelated boundary conditions.

To understand the hydraulic behaviour of a system, it is first necessary to categorise these boundary conditions and to understand their mutual topological and algebraic properties. Use of graph theory is considered to be the best tool for network analysis, which describes the topological and algebraic properties between different elements of the network in the form of matrices.

A pipe network may be described by a directed graph in which all components and hydraulic constraints are either nodes or elements. Hence, all boundary conditions in a network are either nodes or elements in a network graph.

In this chapter, different types of boundary conditions are first divided into six types of elements and a graph-theoretical model of a pipe network is presented. These six types of elements have been used in the mathematical formulation of different network problems considered in this study.

3.1   A Graph-theoretical Model of Pipe Network

Procedures for modelling of piping systems based on graph-theoretical concepts have been extensively considered by various researchers (Chandrashekar and Kesavan, 1972, 1980; Lam and Wolla, 1972; Onizuka, 1986; Shimada, 1989, 1992 among others). These procedures combine two distinct aspects of the system in arriving at the final system of equations i.e. the system structure (topology) and the system component's characteristics. The system structure is explicitly modelled by matrices such as node incidence matrix, loop incidence matrix and cut-set matrix containing 0 or ±1 entries representing the continuity and geometric constraints whereas the component characteristics are obtained either by laboratory experiments or empirical relationships. The well-known Darcy-Wiessbach equation is an example of the empirical relationship between the head loss and the flow rate associated with a pipe.

A pipe network can be considered as a collection of several types of components that are interconnected in a specified manner. A pipe, for example, has two ends called the nodes at which it can be connected to a network. Similarly, a pump can be connected at the two nodes to the network creating a certain flow through it. A pipe network can be represented by a directed connected graph G comprising a finite number of edges or elements connecting nodes pair-wise. All system components are either the elements or nodes of the graph. Interconnections or junctions of pipes and other components are represented by nodes.

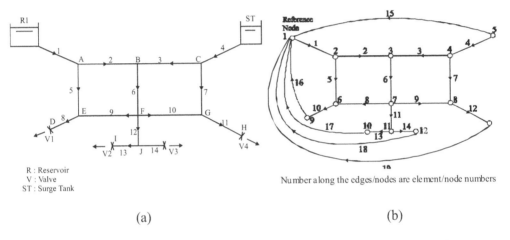

(a)   (b)

Fig. 3.1 (a) A pipe network; (b) Directed graph of pipe network

3.2   Network Boundary Conditions : Types of Elements in a Network Graph

Hydraulic pipe networks consist of several types of components such as reservoir, surge tank, valve, pipe, pump etc. In the directed graph of a network each component is either an element or a node. In the graph, a node is characterised by a particular head, H and an outflow discharge, Q. Similarly, an element is associated with a particular head loss, h and a discharge, q.

Mathematically, a node or an element is equivalent to a boundary where a special boundary condition or specification is imposed. Depending upon whether the head loss and/or discharge variable is a known quantity in an element, possible boundary conditions in an element can be categorised into six categories called, hereinafter, element types. Similarly, depending upon whether the head and/or outflow discharge is a known quantity at a node, possible boundary conditions at a node can be categorised into six categories called, hereinafter, node types. However, by joining all the nodes to a selected reference node of known head with pseudo-elements, all nodal boundary conditions are transformed into element boundary conditions. Joining of nodes with the reference node by pseudo-elements makes an extended network, which is closed. It is important to note that only those nodes, which have network components or a boundary condition, are joined with the reference node. Junction nodes with no boundary condition need not be joined to the reference node.

Hence, an arbitrary network comprising of different types of components or boundary conditions is represented by six types of elements. These six element types are as follows:

Element Type 1 or Pressure Constrained Element

Type 1 or pressure constrained elements are those which have a specified or known head loss and an unknown arbitrary discharge through it. Relationship between head loss and discharge through this element is unknown. Characteristic of type 1 element is shown in Fig. 3.2. Head loss and discharge in this element are represented by h_1^* and q_1 respectively.

In unsteady flow problems, this element has a specified or known time history of head loss, $h_1^*(t)$, and unknown discharge variation, $q_1(t)$, through it. (Asterix (*) denotes the known value of head loss.)

All elements joining reservoir nodes having known heads or nodes with specified head with the datum or reference node are type 1 elements. Network components, which are to be designed for a specified head loss, are also type 1 elements. For example, a pipe with unknown diameter with specified head loss, a valve with unknown opening with specified head loss etc.

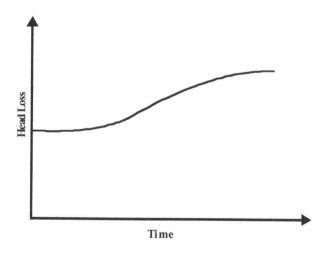

Fig. 3.2 Characteristic of element type 1

Element Type 2 or Flow Constrained Element

Type 2 elements are those which have a specified discharge and an unknown arbitrary head loss through it. In this element also, relationship between head loss and discharge is unknown. Characteristic of this element is shown in Fig. 3.3. Head loss and discharge through this element are represented by h_2 and q_2^* respectively.

In unsteady flow problems, these elements have specified time history of discharge, $q_2^*(t)$, and unknown head loss variation, $h_2(t)$.

All elements joining nodes having known outflows or inflows with the reference node are of type 2. Network components which are to be designed for a specified discharge are type 2 elements such as a pipe with unknown diameter to be designed for a specified discharge, designing a valve opening for the specified discharge through it etc.

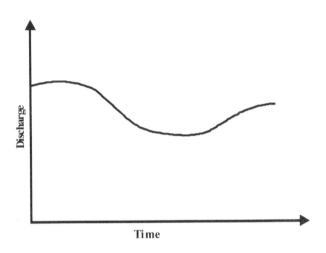

Fig. 3.3 Characteristic of element type 2

Element Type 3 or Pressure-Flow Constrained Element

Type 3 or pressure-flow constrained elements are those which have both specified or known head loss and discharge through it. Relationship between head loss and discharge through this element is, thus, known. Characteristic of type 3 element is shown in Fig. 3.4. Head loss and discharge in this element are represented by h_3^* and q_3^* respectively. In unsteady flow problems, this element has a specified or known time history of both head loss, $h_3^*(t)$, and discharge, $q_3^*(t)$, through it.

All network components with known relationship between head loss and discharge through it and specified head loss or discharge, are of type 3. Elements joining nodes having specified both head and outflow (or inflow) with the reference node are type 3 elements.

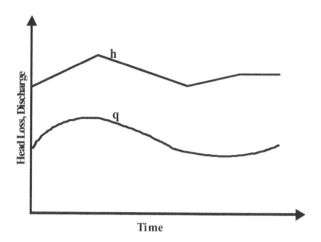

Fig. 3.4 Characteristic of element type 3

Element Type 4 or Open Element

Type 4 or open elements are those that have both unknown head loss and discharge through it. Relationship between head loss and discharge through this element is also unknown. Head loss and discharge in this element are represented by h_4 and q_4 respectively.

In unsteady flow problems, this element has an unspecified or unknown time history of both head loss, $h_4(t)$, and discharge, $q_4(t)$, through it.

Network components, which are to be designed and may have any arbitrary head loss and discharge through it, are of type 4.

Element Type 5

Type 5 elements are those that have both, unknown head loss and unknown discharge through it, though the relationship between head loss and discharge through this element is known. Characteristic of type 5 element is shown in Fig. 3.5. Head loss and discharge in this element are represented by h_5 and q_5 respectively.

In unsteady flow problems, this element has an unknown time history of both head loss, $h_5(t)$, and discharge, $q_5(t)$, through it. Type 5 element can be described through following relation:

$$h_5(t) = h_5(q_5(t)) \quad \text{or} \quad q_5(t) = q_5(h_5(t)) \tag{3.1}$$

All network components with known physical parameters having unknown head loss and discharge are type 5 elements, such as, an existing pipe with known diameter and other physical parameters, a surge tank with known area, a pump with known characteristic curve, a valve with known area of opening etc.

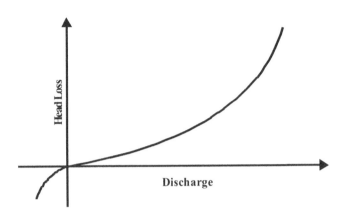

Fig. 3.5 Characteristic of element type 5

Element Type 6 or Ideal Check Valve

Type 6 elements represent ideal check valve conditions. For an ideal check valve if there is a flow through it, there is no head loss and if no flow passes through it, head loss across the valve exists. Hence,

$$q_6 > 0 \ , \ h_6 = 0 \ \text{and when} \ q_6 = 0 \ , \ h_6 < 0 \tag{3.2}$$

and thus,

$$q_6 \, h_6 \ = \ 0 \tag{3.3}$$

Characteristic of element type 6 is shown in Fig. 3.6.

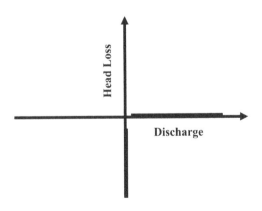

Fig. 3.6 Characteristic of element type 6

3.3   Algebraic and Topological Properties of a Network Graph

Consider a directed graph of a network consisting of different types of elements as described in the previous section. If a network consists of ne number of elements and nn number of nodes, number of fundamental loops in the network are given by

$$nl = ne - nn + 1 \qquad (3.4)$$

If the same types of elements are grouped together, the node incidence matrix, N, of the graph can be written as,

$$N = \begin{bmatrix} N_1 & N_2 & N_3 & N_4 & N_5 & N_6 \end{bmatrix} \qquad (3.5)$$

Similarly, loop incidence matrix, L, after grouping same type of elements, can be written as,

$$L = \begin{bmatrix} L_1 & L_2 & L_3 & L_4 & L_5 & L_6 \end{bmatrix} \qquad (3.6)$$

A detailed description of algebraic and topological properties of a network graph is given in Appendix A. These properties have been used in the mathematical formulation of a network problem.

3.4   Series and Parallel Connections : Equivalent Element

In a network, all series and parallel connections between two elements can be converted into a single equivalent element. The type of equivalent element depends on the type of two elements that make series or parallel connection.

For a series connection of two elements, e1 and e2 as shown in Fig. 3.7, the head loss through equivalent element is the sum of the head losses of two elements, i.e.

$$h_{eq} = h_{e1} + h_{e2} \qquad (3.7)$$

The discharge through the equivalent element will be equal to the discharge in the two elements i.e.,

$$q_{eq} = q_{e1} = q_{e2} \qquad (3.8)$$

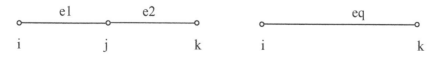

Fig. 3.7 Series connection of two elements and their equivalent element

Considering elements type 1, 2, 3, 4 and 5, possible series connections between different type of elements and their equivalent element are listed in Table 3.1. Some series connections between two types of elements are not allowed because of restrictions imposed by necessary and sufficient conditions for the existence and uniqueness of a solution, which are derived in the next chapter.

Table 3.1 Type of equivalent element for series connection of two elements

Type of Element e1	Type of Element e2	Type of Equivalent element	Remarks
1	1	1	$h_{eq} = h_{e1} + h_{e2}$
1	2	2	$q_{eq} = q_{e2}$
1	3	3	$h_{eq} = h_{e1} + h_{e2}$
			$q_{eq} = q_{e2}$
1	4	4	
1	5	5	
2	2	-	Not allowed
2	3	-	Not allowed
2	4	-	Not allowed
2	5	2	$q_{eq} = q_{e1}$
3	3	-	Not allowed
3	4	2	$q_{eq} = q_{e1}$
3	5	3	$h_{eq} = h_{e1} + h_{e2}\,(q_{e1})$
			$q_{eq} = q_{e1}$
4	4	-	Not allowed
4	5	4	
5	5	5	

Similarly, type of equivalent element depends on the type of elements making parallel connection. For a parallel connection of two elements, e1 and e2, as shown in Fig. 3.8, the head loss and discharge through their equivalent element is given by

Fig. 3.8 Parallel connection of two elements and their equivalent element

$$h_{eq} = h_{e1} = h_{e2} \tag{3.9}$$

$$q_{eq} = q_{e1} + q_{e2} \tag{3.10}$$

Table 3.2 shows the type of equivalent element for the parallel connections of different types of elements. Some parallel connections between two types of elements are not allowed due to the restrictions imposed by the necessary and sufficient conditions, which are derived in next chapter.

Table 3.2 Type of equivalent element for parallel connection of two elements

Type of Element e1	Type of Element e2	Type of Equivalent element	Remarks
1	1	-	Not allowed
1	2	1	$h_{eq} = h_{e1}$
1	3	-	Not allowed
1	4	-	Not allowed
1	5	1	$h_{eq} = h_{e1}$
2	2	2	$q_{eq} = q_{e1} + q_{e2}$
2	3	3	$h_{eq} = h_{e2}$
			$q_{eq} = q_{e1} + q_{e2}$
2	4	4	
2	5	5	
3	3	-	Not allowed
3	4	1	$h_{eq} = h_{e1}$
3	5	3	$h_{eq} = h_{e1}$
4	4	-	Not allowed
4	5	4	
5	5	5	

Transformation of two series or parallel elements into an equivalent element is valid for network models used for steady flow and pressure surges i.e. slow transients only. Moreover, in the case of network model used for pressure surges, parallel connection of two type 5 elements making one equivalent element is not possible. For this case two parallel type 5 elements should be retained.

After transforming series and parallel connections, a network consists of at least three elements at a node. The number of elements in such a network is, hence, given by,

$$ne \geq \frac{3nn}{2} \tag{3.11}$$

The number of loops in such a network, thus, becomes

$$nl \geq \frac{nn}{2} + 1 \tag{3.12}$$

Hereinafter, networks having at least three elements incident at a node have been considered for different problem formulations.

3.5   Definition of Matrices and Vectors

In this section, matrices and vectors related to a graph G of a network are defined. Hydraulic behaviour of a network is represented in terms of head loss and discharge in its components or the nodal heads and discharges in the components. In the network comprising of element type 1, 2, 3, 4 and 5 these three, head loss, nodal head and discharge, vectors are as follows:

Head Loss Vector h

Head loss vector h is defined as

$$h = \{ h_i \} = \{ H_j - H_k \} (i = 1, 2, \dots .ne) \tag{3.13}$$

in which H_j and H_k are heads at the upstream (jth node) and downstream (kth node) ends of an element respectively. If same type of elements are grouped together, head loss vector can be written as

$$h = \begin{Bmatrix} h_1^* \\ h_2 \\ h_3^* \\ h_4 \\ h_5 \\ h_6 \end{Bmatrix} \tag{3.14}$$

h_1^*, h_2, h_3^*, h_4, h_5 and h_6 are sub-vectors for element type 1, 2, 3, 4, 5 and 6 respectively.

Nodal Head Vector H

Nodal head vector H is defined as

$$H = \{ H_i \} \; (i = 1, 2, \dots .nn) \tag{3.15}$$

The nodal head vector H is associated with the head loss vector h and node incidence matrix by the following relationship

$$h = N^T H \qquad (3.16)$$

Discharge Vector q

Discharge vector q is defined as

$$q = \{ q_i \} \ (i = 1, 2, \ldots\ldots ne) \qquad (3.17)$$

where q_i is the discharge in ith element. If the same types of elements are grouped together, discharge vector can be written as

$$q = \begin{Bmatrix} q_1 \\ q_2^* \\ q_3^* \\ q_4 \\ q_5 \\ q_6 \end{Bmatrix} \qquad (3.18)$$

q_1, q_2^*, q_3^*, q_4, q_5 and q_6 are discharge sub-vectors corresponding to elements type 1, 2, 3, 4, 5 and 6 respectively.

3.6   Network Flow Models

Steady state flow

System of equations describing a steady state in a network is based on Kirchhoff's laws. For an arbitrary network consisting of different types of elements, Kirchhoff's node law describing the continuity of flow at nodes can be written as

$$N.q = 0 \qquad (3.19)$$

Kirchhoff's loop law for the network can be written as

$$L.h = 0 \qquad (3.20)$$

Apart from eqs. 3.19 and 3.20, constitutive equations for all the elements type 5 are given by eq. 3.1. For pipe elements, the Darcy-Wiessbach equation describing the relationship between head loss and discharge is given by

$$h = \frac{fl}{2gDA^2} q^2 \qquad (3.21)$$

where f, l, D and A are coefficient of friction, length, diameter and cross sectional area of the pipe respectively. g is acceleration due to gravity.

For pumps, the constitutive equation considering the monotonic behaviour is given by

$$h = a - b.q^2 \qquad (3.22)$$

where a and b are constants.

For flow controlling valves, the orifice equation describing the flow through it is written as

$$q = C_d A_v \sqrt{2gh} \qquad (3.23)$$

where C_d and A_v are coefficient of discharge and area of orifice of the valve.

Equations 3.19-3.23 are the governing equations for the steady state simulation of a network consisting of reservoirs, pipes, valves and pumps. Besides these equations, the specifications for type 1, 2 and 3 elements also constitute the set of equations.

Transient flow analysed using rigid water column theory

The system of equations describing the transient behaviour using rigid water column theory consists of Kirchhoff's laws as given by eqs. 3.19 and 3.20 and equations of motion for pipes and surge tanks.

The equation of motion for pipes considering one dimensional flow can be written as (Wylie, 1993)

$$h = m \frac{dq}{dt} + r q |q| \qquad (3.24)$$

where $m = l/(gA)$ and $r = f.l/(2gDA^2)$.

The equation of motion for surge tanks is given by

$$q_s = A_s \frac{dH_s}{dt} \qquad (3.25)$$

where A_s is the area of surge tank and H_s is the water level in the surge tank.

In this study, the dynamic behaviour of control valves is neglected and the constitutive equation for valves remains same as for steady state flow.

Transient flow analysed using elastic water column theory

For transients analysed using elastic water column theory, the equation of motion and continuity in the pipes for one dimensional flow and neglecting convective term can be written respectively as follows (Wylie, 1993)

$$\frac{\partial q}{\partial t} + g A \frac{\partial H}{\partial x} + f \frac{q|q|}{2DA} = 0 \tag{3.26}$$

$$\frac{gA}{a^2} \frac{\partial H}{\partial t} + \frac{\partial q}{\partial x} = 0 \tag{3.27}$$

where a is the wave speed. In networks, besides the above equations for each pipe, continuity equation and condition of common head at junctions are to be satisfied. For valves and surge tanks, additional equations given by 3.23 and 3.25 are used.

3.7   Concluding Remarks

Hydraulic behaviour of a pipe networks whether in steady state or in unsteady state, is described by discharges and head losses in its components. In the presence of several types of components, network configuration becomes complex and hence, mathematical formulation of a network problem is not straightforward. An efficient network modelling takes care of this problem. Network modelling comprises of mainly three steps, simulation of network components, mathematical formulation of a well-posed problem and an efficient solution algorithm to the problem. In this chapter a generalized network model is presented using graph-theoretical concepts.

The use of algebraic topology and matrix algebra in the study of hydraulic network theory has been written about by many authors and is now well established. Algebraic topology and matrix algebra together provides an extremely useful mathematical language for network theory. Use of graph theory is helpful not only in providing a simplified mathematical model of a problem but also in showing relationships between different types of network components.

In this chapter, a graphical representation of an arbitrary pipe network is developed. Network components are either nodes or elements of a graph. Possible network components have been categorised in six element types. By joining nodes with the reference node, all nodal components are also incorporated within these six element types. As hydraulic behaviour of a network is well described by discharge and head loss in its components, this division of elements into six types takes care of different possible combinations of head loss and discharge that may be present in a network. Mathematically, these six element types are six boundary conditions.

A network problem whether it is steady state or unsteady state, may comprise of some or all types of boundary conditions. Hence, a network of arbitrary configuration comprising of different types of components is a set of six types of elements interconnected in a predefined manner.

Finally, some algebraic and topological properties of the network graph and associated matrices and vectors have been defined which will be used in the formulation and solution of different types of problems.

4 | Network Solvability

The unknown parameters in a pipe network can be element head losses, element flows and element resistances. For network analysis to be feasible, the number of unknown parameters and their distribution over the entire network must be such that an adequate number of independent equations can be formulated. If the unknown parameters are concentrated in one part of the network and known parameters in the other, it is not possible to determine all the unknown parameters uniquely. Therefore, the number and distribution of the unknown parameters should obey certain rules so that the network analysis is feasible. Determination of these rules is necessary for the meaningful and effective modelling of network problems.

In this chapter, first a critical review of past solvability rules is done and, then, necessary and sufficient conditions for the existence and uniqueness of a solution for a arbitrary network comprising of different types of elements are developed.

4.1   A Critical Review of Past Solvability Rules

Shamir and Howard (1968) may have been the first to consider the problem of solvability of pipe networks. They suggested some rules for the existence and uniqueness of solution of networks. These rules (rules no. 1) are as follows:

(1)   The total number of the unknowns must be equal to the total number of the nodes in the network.
(2)   At least one nodal head must be known.
(3)   A node having an unknown consumption must be connected to at least one other node with a known consumption.
(4)   A subsystem consisting of a pipe with unknown resistance and its two end nodes, must have no more than one additional unknown – one of the heads or consumptions at the two end nodes.
(5)   Considering any node, at least one of the following must be unknown : the consumption at the node, the head at the node itself or at any adjacent node, or the resistance of a pipe connected at the node.

These rules were developed heuristically by considering different combinations of unknowns, and no rigorous mathematical proof was provided for their validity. This set of conditions is necessary to guarantee that the jacobian matrix for the nodal method of analysis is non-singular.

These rules lack generality in the sense that they do not incorporate elements type 1, 2 and 3. Though nodes type 1, 2 and 3, generating pseudo-elements type 1, 2 and 3 respectively, are considered while driving these rules, it is observed that presence of elements type 1, 2 and 3 in the network makes the system more complicated. Moreover, it is found that even for simple networks these rules are incorrect and insufficient. Consider a pipe network (a) as shown in Fig. 4.1 consisting of four nodes with specified heads and unknown outflows. This network violates rule 3 but possesses a unique solution. Flow in each pipe can be uniquely determined by knowing heads at its two ends and, then, continuity equations at each node uniquely defines the outflow.

As an another example, consider a network (b) shown in Fig. 4.1. This network violates rule 4 of Shamir and Howard as the pipe with unknown resistance has two unknowns at its two end nodes. However, the network possesses a unique solution. Flow in pipe between nodes type 3 and 1 is uniquely determined knowing the head loss. From continuity at node type 3, flow and, thus, head loss in pipe between node type 3 and 2 is uniquely determined. This gives head at node type 2 uniquely. From continuity at node type 2, flow in pipe type 4 is uniquely determined and the difference of head at node type 2 and 1 gives the head loss in this pipe.

In general, it is observed that rules presented by Shamir and Howard are incorrect, insufficient and are not general. For a network consisting of different types of elements as described in previous chapter, these rules fails to provide necessary and sufficient conditions for the existence and uniqueness of a solution.

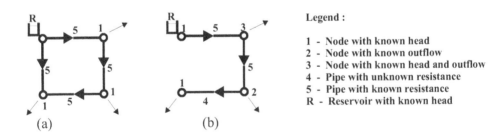

Fig. 3.1 Networks violating solvability rules no. 1 but having unique solutions

Using graph theory, Gofman and Rodeh (1972) also developed certain rules for solvability of networks. They termed pipes with unknown characteristics as head generators. Their rules (rules no. 2) are as follows :

(1)   The number of nodes with known heads must exceed the number of head generators by one.

(2)   For a given network G, there exists a connected network G' such that:

a. All head generators and nodes with known heads belong to G'.
b. G' is loopless.
c. A path in G' connecting two nodes of known nodal heads contains at least one head generator.

These rules are also not correct and lacks generality as that of rules presented by Shamir and Howard. Consider a pipe network (a) as shown in Fig. 4.2. This network violates the above rules in the sense that there is a loop made by nodes with known head and head generator. However, network possesses unique solution. Flows in elements type 5 are uniquely determined by knowing heads at its two ends and continuity equations at nodes uniquely defines outflows and flow in element type 4.

Consider another example of network (b) as shown in Fig. 4.2. This network obeys all the rules of Gofman and Rodeh but does not possess a unique solution. Knowing the head loss in type 3 element and head at its upstream node, head at node type 2 is uniquely determined which in turn uniquely defines the flow in pipe connecting this node with reservoir node. This makes the case for possible violation of continuity at node type 2 unless the sum of the flow in the connecting pipes is equal to the specified outflow. In any case, flow in other two pipes can not be determined.

Rules presented by Gofman and Rodeh are also not correct and complete.

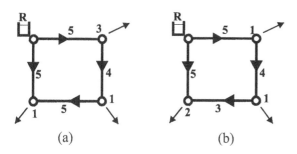

Legend :

1 - Node with known head
2 - Node with known outflow
3 - Node/pipe with known head/
head loss and outflow/flow
4 - Pipe with unknown resistance
5 - Pipe with known resistance
R - Reservoir with known head

(a)   (b)

*Fig. 4.2 (a) Network violating solvability rules no. 2 but having unique solution
(b) Network obeying solvability rules no. 2 but possess no solution*

Network solvability rules are also presented by Bhave (1990). These rules were also developed heuristically without any proof and are as follows (rules no. 3):

(1)   The total number of unknowns must be equal to the total number of nodes in the network.
(2)   At least one nodal head must be known and one nodal flow unknown.
(3)   Considering any node, at least one of the following must be unknown : the flow at the node, the head at the node itself or at any adjacent node, or the resistance of the pipe connected to the node.
(4)   Each pipe with an unknown resistance must lie on an independent path connecting two nodes of known heads, and each such path must not contain more than one pipe with an unknown resistance. All these paths must form branching configurations without any loops.

Network (b) in Fig. 4.2 obeys the rules presented by Bhave but as shown earlier do not possess a unique solution. Network (a) in Fig. 4.3 also obeys Bhave's rules but it can be shown that it possess non-unique solution. Network (b) in Fig. 4.3 also possess non-unique solution for some combination of pipe resistances. In network (b) of Fig. 4.3, if resistances of all the four type 5 pipes are same, it will have infinite solution.

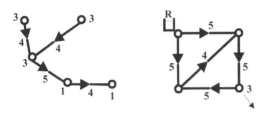

Legend :

1 - Node with known head
2 - Node with known outflow
3 - Node with known head and outflow
4 - Pipe with unknown resistance
5 - Pipe with known resistance
R - Reservoir with known head

Fig. 4.3 Networks obeying rules no. 3 but having non-unique solutions

In general, it is observed that solvability rules present in the literature are not correct, incomplete and lacks generality and, thus, do not guarantee a unique solution.

Determination of necessary and sufficient conditions for the existence and uniqueness of a solution is must for effective modelling of pipe networks.   In the next section, these conditions have been derived for an arbitrary network consisting of elements type 1, 2, 3, 4 and 5.

4.2   Theorems Defining Necessary and Sufficient Conditions for the Existence and Uniqueness of a Solution

For an arbitrary network comprising of different elements as described in chapter 2, a unique solution of a network problem may exist only if the following necessary and sufficient conditions are satisfied.

First Necessary Condition

Theorem 4.1 *: In a network, at least one nodal head should be known.*

Proof : The known nodal head serves as the reference node, relative to which other nodal heads are determined. If a network consists of all unknown nodal heads, all element head losses can be determined but nodal heads can not be defined uniquely.

Second Necessary Condition

Theorem 4.2 *: In a network, number of element type 3, ne3, should be equal to the number of element type 4, ne4.*

Proof : The number of conditions, nc, to be satisfied by a mathematical model are Kirchhoff's nodal law, Kirchhoff's loop law, known head loss in element type 1, known flow through element type 2, known head loss and flow through element type 3 and known relationship between head loss and flow in element type 5 i.e.

$$nc = nn - 1 + nl + ne1 + ne2 + 2\ ne3 + ne5$$
$$= ne + ne1 + ne2 + 2\ ne3 + ne5 \tag{4.1}$$

The total number of variables, nv, are

$$nv = 2\ ne \tag{4.2}$$

i.e. flows and head losses through all elements.

For the solvability of the network, number of variables must be matched by the number of conditions. From eqs. 4.1 and 4.2 it immediately follows that :

$$ne3 = ne4 \tag{4.3}$$

Third Necessary Condition

Theorem 4.3 : *No cut of the network may be formed by elements type 2 and/or 3 only.*

Proof : In fact, the flows through the elements of any arbitrary cut, such as cut C in Fig. 4.4 (a) are not independent since they must satisfy the gross continuity condition of the sub-networks obtained by removing the elements of the cut from the original network. Therefore if all the flows through the cut are specified as in Fig. 4.4 (a), either the gross continuity condition is violated or one of the specifications is redundant. In the former case no distribution of flows satisfying the node law can be found and, thus, no solution exists. In the latter case the problem is underdetermined. With reference to Fig. 4.4 if :

a. $q_1 + q_2 + q_6 \neq 0$, then the continuity condition can not be satisfied at node 1 and no solution exists.
b. $q_1 + q_2 + q_6 = 0$, the continuity condition is identically satisfied at node 1 and therefore this condition can be dropped from the model. The resulting system has nv = 18 and nc = 17 and is underdetermined.

It is worth noting that this necessary condition also applies to the cut separating the reference node from the rest of the network.

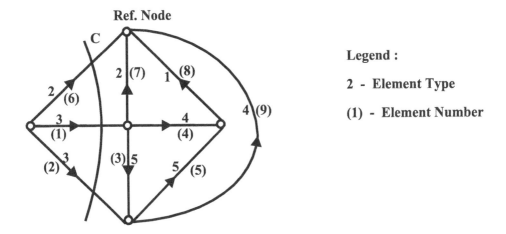

Fig. 4.4 Network violating the third necessary condition

Fourth Necessary Condition

Theorem 4.4 : *No loop of the network may be formed by elements type 1 and/or 3 only.*

Proof : By a similar reasoning to the one used for the third condition it can be concluded that either Kirchhoff's second law can not be satisfied and no solution exists or one of the specifications is redundant and the system is underdetermined. For the loop l in Fig. 4.5 ;

a. either $h_5 + h_8 - h_9 \neq 0$ and no solution exists.
b. or $h_5 + h_8 - h_9 = 0$, $nv = 18$, $nc = 17$ and the system is underdetermined.

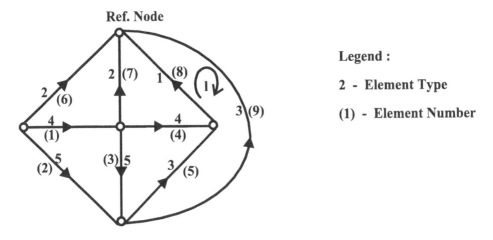

Fig. 4.5 Network violating the fourth necessary condition

Fifth Necessary Condition

Theorem 4.5 : *No loop of the network may be formed by elements type 1 and/or 4 only.*

Proof : If a loop of elements type 1 and/or 4 exists, any arbitrary loop flow will satisfy all nodal and loop conditions. Hence, the system will not have a unique solution and, therefore, contains infinite number of solutions.

The flows through the elements 5, 8 , 9 which form the loop 1 in Fig. 4.6 can not be determined uniquely. Assuming that q_5^1, h_5^1, q_8^1, h_8^1, q_9^1, h_9^1 is a solution then so is :

$$q_5^2 = q_5^1 + \beta l \qquad\qquad h_5^2 = h_5^1$$
$$q_8^2 = q_8^1 + \beta l$$
$$q_9^2 = q_9^1 + \beta l \qquad\qquad h_9^2 = h_9^1 \qquad\qquad (4.4)$$

The above expressions produce an infinity of solutions provided βl is maintained in a suitable interval such that there is no flow reversal in any of the elements type 4 forming the loop.

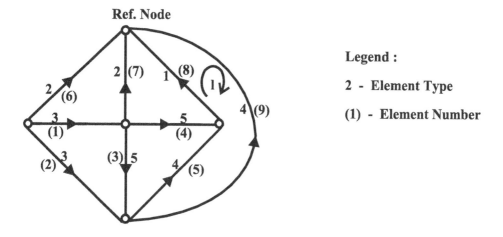

Fig. 4.6 Network violating the fifth necessary condition

Sixth Necessary Condition

Theorem 4.6 : *No cut of the network may be formed by elements type 2 and/or 4 only.*

Proof : If a cut of elements type 2 and/or 4 exists, any arbitrary head loss through these elements will satisfy all nodal and loop conditions. A cut of elements type 2 and/or 4 divides a network in two sub networks connected by elements that can pusses any arbitrary head loss. Hence, the system will not have a unique solution and, therefore, contains infinite number of solutions.

The head loss across the elements 1,2 and 6 in Fig. 4.7 can not be uniquely determined. Assuming that $q_1^1, h_1^1, q_2^1, h_2^1, q_6^1, h_6^1$ is a solution, then so is :

$$\begin{array}{ll} h_1^2 = h_1^1 + Z_1 & q_1^2 = q_1^1 \\ h_2^2 = h_2^1 + Z_1 & q_2^2 = q_2^1 \\ h_6^2 = h_6^1 + Z_1 & q_6^2 = q_6^1 \end{array} \qquad (4.5)$$

The above expression exhibit an infinite number of solutions provided the arbitrary parameter Z_1 is maintained in an interval such that no sign change in the head losses across any of the elements type 4 occurs.

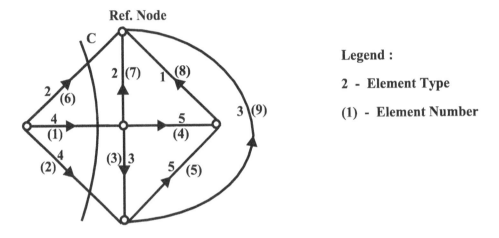

Fig. 4.7 Network violating the sixth necessary condition

4.2.1 Implications of the Six Necessary Conditions

The six necessary conditions developed above imply a number of properties related to the connectivity of the various types of elements. These properties are :

Property 1 *: In a network, number of elements type 3 can not be more than number of branches or chords whichever are less.*

This follows directly from conditions 3 and 4 which prohibit a cut and loop of elements type 3.

Property 2 *: In a network, number of elements type 4 can not be more than number of branches or chords whichever are less.*

This follows directly from conditions 5 and 6 which prohibit a loop and cut of elements type 4.

Property 3 : *In a network, sum of elements type 1 and 3 can not be more than number of branches.*

It follows directly from condition 4.

Property 4 : *In a network, sum of elements type 2 and 3 can not be more than number of chords.*

It follows directly from condition 3.

Property 5 : *In a network, sum of elements type 1 and 4 can not be more than number of branches.*

It follows directly from condition 5.

Property 6 : *In a network, sum of elements type 2 and 4 can not be more than number of chords.*

It follows directly from condition 6.

Property 7 : *Elements type 1 and 3 are branches of some tree T_{13} of the network.*

This follows directly from condition 4. It is evident that the property also applies to elements type 1 and 3 separately.

Property 8 : *Elements type 1 and 4 are branches of some T_{14} of the network.*

This follows directly from condition 5. The property also applies to elements type 4 separately.

Property 9 : *There exists at least one spanning tree T_{14} such that :*
a.   All elements type 1 and 4 are branches of T_{14}.
b.   All elements type 2 and 3 are chords of T_{14}.

The proof is by construction. From the necessary condition 3 it follows that the network obtained by deletion of all the elements type 2 and 3 is connected and, therefore, that at least one spanning tree T which excludes all elements type 2 and 3 exists. If T is unique, it necessarily contains the tree T_{14} of Property 8 . Otherwise, at least one spanning tree which includes T_{14} may be found. In either case T is composed of the trees of T_{14} and paths of elements type 5 joining these trees. Finally, T satisfies a. and b. and is the spanning tree T_{14}.

Property 9 implies the existence of a base of flow variables which includes all the flows through elements type 1 and 4. Such a base is formed by the flows through the branches of the tree T_{14}. It is a well-known fact that these flows can be uniquely determined from the flows through the chords of T_{14}.

Property 10 : *There exists at least one spanning tree T_{13} such that*
a.   All elements type 1 and 3 are branches of T_{13}.

b.   *All elements type 2 and 4 are chords of T_{13}.*

The proof is similar to the one given above and follows from condition 6 and property 7.

Property 10 implies the existence of a base of head loss variables which includes all the head losses through elements type 2 and 4. Such a base is formed by the head losses across the chords of tree T_{13}. It is a well-known fact that these can be determined uniquely from the head losses across the branches of T_{13}.

Seventh Necessary Condition

Theorem 4.7 : *In a network, there exists at least one tree T_{13a} and at least one T_{14a} having the following properties :*

a.   *T_{13a} is a spanning tree. Its branches include all elements type 1 & 3 and a subset of elements type 5 (called 'a'). The co-tree of T_{13a} includes all elements type 2 & 4 and the remaining elements type 5 (called 'b').*
b.   *T_{14a} is a spanning tree. Its branches include all elements type 1 & 4 and same subset of elements type 5 (called 'a') as included in spanning tree T_{13a}. The co-tree of T_{14a} includes all elements type 2 & 3 and the same elements type 5 (called 'b') as included in the co-tree of T_{13a}.*

Proof : The proof the theorem is based on the following Corollaries :

Corollary 4.7.1 : If spanning trees T_{13a} and T_{14a} exist, each element type 4 contains at least one element type 3 in its loop and vice versa.

Proof : Consider a spanning tree T_{13a} in a network and extract all elements type 2, 4 and b. Addition of one element type 4 would create a fundamental loop and would include elements type 1, 3 and/or a as branches. If the loop made by element type 4 dose not include any element type 3, a loop of elements type 1, 4 and a exists and, thus, a spanning tree T_{14a} dose not exist. Hence, if spanning tress T_{13a} and T_{14a} exist, each element type 4 contains at least one element type 3 in its loop.

Similarly, it can be proved that each element type 3 contains at least one element type 4 in its loop when spanning tree T_{14a} is considered.

Corollary 4.7.2 : If spanning trees T_{13a} and T_{14a} exist, no two elements type 4 contains the same elements type 3 in their loops and vice versa.

Proof : Consider a spanning tree T_{13a}. Extract all elements type 2, 4 and b. Addition of a element type 4 would create a loop which will contain elements type 1, 3 and a. Contract all elements type 1 and a included in the loop. This creates a loop of element type 4 and at least one element type 3 only. Add another element type 4. This element type 4 makes another loop with elements type 1, 3 and a. Contract all elements type 1 and a of this loop. Thus, the loop made by this element type 4 contains only elements type 3. If the two elements type 4 contains the same elements type 3 in their respective loops, extraction of these elements type

3 would create a loop of these two elements type 4. This implies the non-existence of spanning tree T_{14a}. Hence, if spanning trees T_{13a} and T_{14a} exist, no two elements type 4 contains the same elements type 3 in their loops

Similarly, it can be proved that no two elements type 3 contain the same elements type 4 in their loops if both spanning trees T_{13a} and T_{14a} exist.

Corollary 4.7.3 : If spanning trees T_{13a} and T_{14a} exist, no element type 4 contains in its loop only those all elements type 3 which are not common in the loops made by any other two elements type 4 and vice versa i.e. if an element type 4 contains in its loop a set of elements type 3, called x, and if another element type 4 contains a set of elements type 3 in its loop, called y, no other element type 4 contains in its loop only those elements type 3 which are included in x ∩ y.

Proof : The proof is similar to the proof of previous Corollary. Consider a spanning tree T_{13a}. Extract all elements type 2, 4 and b. Add two type 4 elements and contract all elements type 1 and a included in their loops. These two type 4 elements do not contain in their loops all the same elements type 3 as per Corollary 4.7.2. Add another type 4 element and contract all type 1 and a elements of its loop. If this type 4 element contains only all those type 3 elements which are included by the first two type 4 elements but do not belong to a common set, extraction of all the type 3 elements included by these type 4 elements in their loops will make a loop of type 4 elements. This implies non existence of spanning tree T_{14a}. Hence, if spanning trees T_{13a} and T_{14a} exist, no element type 4 contains in its loop only all those elements type 3 which are included by any other two type 4 elements in their loops but do not belong to a common set.

Similarly, it can be proved if spanning trees T_{13a} and T_{14a} exist, no element type 3 contains in its loop only all those elements type 4 which are included by any other two type 3 elements in their loops but do not belong to a common set.

Corollary 4.7.4 : If spanning trees T_{13a} and T_{14a} exist, a non-singular matrix L_{43} exists for spanning tree T_{13a} and a non-singular matrix L_{34} exists for spanning tree T_{14a}.

Proof: Consider a spanning tree T_{13a} and a loop incidence matrix L_{43} made by type 4 elements as chords with type 3 elements as branches. From second necessary condition given by Theorem 4.2, L_{43} is a square matrix as ne3 = ne4. From Corollary 4.7.1, each row will have at least one non-zero entry. From Corollary 4.7.2, no two rows will have the same entries. This implies that any mathematical operation (addition/subtraction) between two rows will not result in a row with all zero entries. Similarly, from Corollary 4.7.3, it implies that no row will have non-zero entries only in those columns, which may result from addition or subtraction of any other two rows. In other words, any mathematical operation between two rows will not produce a row similar to any other row. From the above argument, the matrix L_{43} has no row with all zero entries and no row can be transformed (by means of matrix transformation i.e. addition/subtraction of one row with another etc.) into a row with all zero entries. This implies that matrix L_{43} has a rank equal to the size of the matrix i.e. ne3. As the rank of L_{43} is equal to its size, the matrix L_{43} is non-singular. Hence, if spanning trees T_{13a} and T_{14a} exist, a non-singular matrix L_{43} exist for spanning tree T_{13a}.

Similarly, it can be proved if spanning trees T_{13a} and T_{14a} exist, a non-singular matrix L_{34} exist for spanning tree T_{14a}.

Corollary 4.7.5 : If matrices L_{43} and L_{34} are singular, spanning trees T_{13a} and T_{14a} do not exist.

Proof: If matrix L_{43} is singular, it means the rank of the matrix is less than ne3. This implies that at least one row of L_{43} has all zero entries or it can be transformed into a row with all zero entries by matrix transformations. This can happen if at least one type 4 element does not include any type 3 element in its loop or if at least one type 4 element possesses the same set of type 3 elements in its loop as included by some other type 4 element or if a type 4 element includes only all those type 3 elements which do not belong to a common set of type 3 elements included by any other two type 4 elements in their loops. In either case, from the arguments presented in Corollary 4.7.1, 4.7.2 and 4.7.3, it can be proved that spanning tree T_{14a} will not exist.

Similarly, it can be proved that if matrix L_{34} is singular, spanning tree T_{13a} will not exist. Hence, if matrices L_{43} and L_{34} are singular, spanning trees T_{13a} and T_{14a} do not exist.

Corollary 4.7.6 : It is necessary that in a network, non-singular matrices L_{43} and L_{34} exist.

Proof: Consider a network with type 1, 2, 3, 4 and 5 elements. Select a spanning tree with all type 1 and 3 elements and some type 5 elements as branches. All type 2 and 4 elements with the remaining type 5 elements are chords. Extract all type 2 and type 5 chord elements and contract all type 1 and type 5 branch elements. The resulting network consists type 3 and 4 elements only. From Property 7 in section 4.2.1, type 3 elements constitute a spanning tree and, thus, all type 4 elements are chords. Network continuity and loop head loss equations using eqs. 3.20 and 3.22 can be written as

$$q_3^* = L_{43}^T q_4 \tag{4.6}$$

$$h_4 + L_{43} h_3^* = 0 \tag{4.7}$$

From eqs. 4.6 and 4.7, q_4 can be evaluated only if the matrix L_{43} is non-singular. Similarly, select a spanning tree with all type 1 and 4 and some type 5 elements as branches. All type 2 and 3 elements and the remaining type 5 elements are chords. Extract all type 2 and type 5 chord elements and contract all type 1 and type 5 branch elements. Resulting network is made of type 3 and 4 elements. From Property 8 in section 4.2.1, type 4 elements constitute a spanning tree and all type 3 elements are chords. Writing network continuity and loop head equations as

$$q_4 = L_{34}^T q_3^* \quad : \quad h_3^* + L_{34} h_4 = 0 \tag{4.8}$$

From eq. 4.8, h_4 can be evaluated only if the matrix L_{34} is non-singular. Hence, in a network, the non-singular matrices L_{43} and L_{34} must exist.

From Corollary 4.7.6, existence of non-singular matrices L_{43} and L_{34} is necessary. From Corollary 4.7.4, if spanning trees T_{13a} and T_{14a} exist, non-singular L_{43} and L_{34} matrices exist. From Corollary 4.7.5, if non-singular matrices L_{43} and L_{34} do not exist, spanning trees T_{13a} and T_{14a} do not exist. Hence, it is necessary that in a network, there exist at least one spanning tree T_{13a} and at least one spanning tree T_{14a}. This proves the Theorem 4.7.

Fig. 4.8 shows a network that violates seventh necessary condition. Network obeys first six necessary conditions but two spanning trees, T_{13a} and T_{14a}, as defined by theorem 4.7 do not exist.

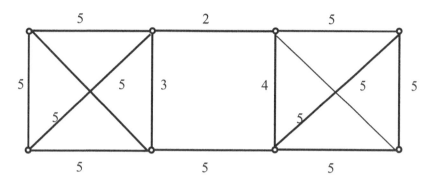

(number along the edges are element type)

Fig. 4.8 Network violating seventh necessary condition

4.2.2 Network Equations Based on Spanning Trees T_{13a} and T_{14a}

Consider a spanning tree T_{13a} in a extended network which contains chords made by elements type 2, 4 and b. The part of loop incidence matrix, L, for the tree elements after grouping together same type of elements can be written as,

$$L = \begin{bmatrix} L_{21} & L_{23} & L_{2b} \\ L_{41} & L_{43} & L_{4b} \\ L_{b1} & L_{b3} & L_{ba} \end{bmatrix} \tag{4.9}$$

`where L_{21} means loop made by elements type 2 with elements type 1 and so on. Similarly, for spanning tree T_{14a} , part of the loop incidence matrix, L', for tree elements can be written as,

$$L' = \begin{bmatrix} L'_{21} & L'_{24} & L'_{2b} \\ L'_{31} & L'_{34} & L'_{3b} \\ L'_{b1} & L'_{b4} & L'_{ba} \end{bmatrix} \tag{4.10}$$

System of equations based on spanning tree T_{13a}

Using equation 3.10 and considering spanning tree T_{13a}, nodal continuity equations of the network can be written as,

$$q_1 = L_{21}^T q_2^* + L_{41}^T q_4 + L_{b1}^T q_b \tag{4.11}$$

$$q_3^* = L_{23}^T q_2^* + L_{43}^T q_4 + L_{b3}^T q_b \tag{4.12}$$

$$q_a = L_{2a}^T q_2^* + L_{4a}^T q_4 + L_{ba}^T q_b \tag{4.13}$$

Loop head loss equations for the network can be written as,

$$h_2 + L_{21} h_1^* + L_{23} h_3^* + L_{2a} h_a = 0 \tag{4.14}$$

$$h_4 + L_{41} h_1^* + L_{43} h_3^* + L_{4a} h_a = 0 \tag{4.15}$$

$$h_b + L_{b1} h_1^* + L_{b3} h_3^* + L_{ba} h_a = 0 \tag{4.16}$$

System of equations based on spanning tree T_{14a}

Considering spanning tree T_{14a}, nodal continuity equations and loop head loss equations can be written as,

$$q_1 = L_{21}'^T q_2^* + L_{31}'^T q_3^* + L_{b1}'^T q_b \tag{4.17}$$

$$q_4 = L_{24}'^T q_2^* + L_{34}'^T q_3^* + L_{b4}'^T q_b \tag{4.18}$$

$$q_a = L_{2a}'^T q_2^* + L_{3a}'^T q_3^* + L_{ba}'^T q_b \tag{4.19}$$

$$h_2 + L_{21}' h_1^* + L_{24}' h_4 + L_{2a}' h_a = 0 \tag{4.20}$$

$$h_3^* + L_{31}' h_1^* + L_{34}' h_4 + L_{3a}' h_a = 0 \tag{4.21}$$

$$h_b + L_{b1}' h_1^* + L_{b4}' h_4 + L_{ba}' h_a = 0 \tag{4.22}$$

Consider continuity equations 4.17-4.19 for spanning tree T_{14a}, and loop head loss equations 4.14-4.16 for spanning tree T_{13a}. From continuity eqs. 4.17 and 4.18, q_1 and q_4 can be determined directly once q_b is known. Similarly, from loop head loss eqs. 4.14 and 4.15, h_2 and h_4 can be determined directly once h_a is known. From, continuity eq. 4.19, q_a can be determined once q_b is known. Hence, q_b is the independent variable and all others are dependent variables.

Necessary and Sufficient Condition

Theorem 4.8 : *In the network, matrix M is non-singular. Matrix M is given by*

$$M = [r_b + L_{ba} r_a L_{ba}^T]$$

(4.23)

Proof. Consider loop head loss equations based on spanning tree T_{13a}, given by eqs. 4.14, 4.15 and 4.16, and continuity equations based on spanning tree T_{14a}, given by eqs. 4.17, 4.18 and 4.19. The only independent variable is q_b. Variables q_1, q_4, q_a, h_2 and h_4 are determined directly from eqs. 4.17, 4.18, 4.19, 4.14 and 4.15 respectively once q_b is known.

Writing eq. 4.16 in terms of flow variables gives,

$$h_b(r_b q_b) + L_{ba} h_a(r_a q_a) = C^*$$

(4.24)

where C^* is a constant given by constant terms of eq. 4.16.

Inserting eq. 4.19 for q_a in eq. 4.24 yields

$$h_b(r_b q_b) + L_{ba} h_a(r_a q_a(q_b)) = C'^*$$

(4.25)

where C'^* is a constant based on constant terms of eqs. 4.16 and 4.19. Equation 4.25 is non-linear and has the only variable q_b.

Equation 4.25 is solved iteratively for q_b. For an iterative solution, the value of q_b for the i+1 iteration is given by

$$q_b^{i+1} = q_b^i - J^{-1} f(q_b^i)$$

(4.26)

where matrix J is given by

$$J = \frac{\partial f}{\partial q_b}\bigg|_{q_b = q_b^i}$$

(4.26)

and f is a function of q_b given by eq. 4.25.

For the solution of eq. 4.26, matrix J should be non-singular. This requires matrix $[r_b + L_{ba} r_a L'_{ba}{}^T]$ should be non-singular.

Hence, it is necessary that in the network matrix M should be non-singular.

If matrix M is non-singular, the variable q_b can be evaluated uniquely. Once q_b is known, all other variables can be evaluated uniquely. Hence, it is sufficient for the existence and uniqueness of a solution of the set of equations given by eqs. 4.14-4.19 that matrix M is non-singular.

4.3   Concluding Remarks

The synthesis approach as applied to analysis, design and operation of networks involves determination of system parameters and flow distribution for the specified head losses and flows in some parts of the network. Different types of specifications along with hydraulic constraints in terms of head losses and discharges are grouped into six elements type as described in previous chapter. Mathematical formulation of a network flow problem whether steady or unsteady in the presence of these six types of elements is not straightforward. Solvability of the resulting models depends on the manner in which these boundary conditions are distributed in the network. Development of these rules in terms of necessary and sufficient conditions is necessary for the effective and meaningful modelling.

Though some solvability rules are present in the literature, they are not complete and are even not correct. Moreover, these rules lack generality in the sense that not all the possible boundary conditions have been incorporated. In this chapter, solvability rules in terms of necessary and sufficient conditions for an arbitrary network consisting of different types of elements have been developed.

The seven necessary conditions and one necessary and sufficient condition can be summarised as one nodal head known, $ne_3 = ne_4$, no cut 2 & 3, no cut 2 & 4, no loop 1 & 3, no loop 1 & 4, existence of T_{13a} & T_{14a} and non-singular matrix M ($=r_b+L_{ba}r_aL'_{ba}{}^T$). It is to point out here that if the seventh necessary condition is satisfied fourth and fifth necessary conditions that state no loop 1 & 3 and no loop 1 & 4 get automatically satisfied. However, these conditions have been retained for better understanding and clarity.

One of the main results in the development of solvability rules is the existence of two spanning trees, T_{13a} and T_{14a}, in the network. Existence of these two spanning trees is a necessary condition but helps in providing efficient algorithms for network flow problems as shown in next chapters. In fact, existence of spanning trees, T_{13a} and T_{14a}, is the base for solving network problems from synthesis approach.

5 | Steady State Simulation : An Optimization Approach

In this chapter, network steady state problems have been divided into two types namely analysis problem and analysis-design problem and a general form of these problems are considered. The proposed methodology for the solution of these problems utilises the concepts of content and co-content developed by Millar for electrical networks including non-linearity.

The major results of this chapter is to show that, provided data and unknowns satisfy conditions necessary for the unique solution of the problem, the resulting steady state distribution of heads and flows may be found without evaluation of the unknown element parameters. Moreover, optimization models developed for the solution of these problems are strictly convex and, thus, provide optimal solutions with guaranteed convergence. Among other advantages, the algorithm developed provides automatic separation of independent and dependent variables and a minimal set of independent state equations. These advantages are particularly significant when the number of unknown parameters is large.

5.1   Analysis and Analysis-Design Problems

Network flow problems can be divided into two types on the basis whether or not element parameters are specified. These are designated as analysis problem and analysis-design problem.

In network analysis problem, the system's hydraulic behaviour i.e. flow and head loss in each element is to be determined for specified pipe system characteristics as well as demand loading and operating conditions. Determination of flow rate in pipes, heads at junctions etc. in the existing distribution network in which all the pipe-system characteristics are known, corresponds to the analysis problem.

The network analysis problem consists of element types 1, 2, 5 and 6 only. When the element type 1 & 2 are pressure-constrained and flow-constrained physical components in the network and not pseudo elements, the definition of the analysis problem can be extended to include the design of these elements.

In analysis-design problems, all the six types of elements are present. It is different from the analysis problem in the sense that the network of analysis-design problem contains some pressure-flow-constrained (type 3) and open (type 4) elements also. Design of new components in the extension and expansion of existing distribution networks in which a required flow-rate & pressure head has to be maintained at some locations corresponds to the analysis-design problem. Hence, in this problem, in addition to the usual unknowns of network analysis, heads and flows, a number of system parameters such as pipe diameters, regulator coefficients etc. are to be determined. Methods for the solution of such problems are of importance for the direct design of parts of the network as well as for sensitivity studies.

5.2   Hydraulic Content and Co-content

The concepts of content and co-content defined by Millar (1951) for solving electrical networks in terms of quantities i (intensity of current) and v (voltage) can easily be extended to hydraulic systems by analogy. The definitions given below are appropriate for hydraulic networks.

Consider an arbitrary element k specified by it's characteristics $h_k = h_k (q_k)$ or $q_k = q_k (h_k)$, where q_k is the flow rate and h_k the corresponding head loss. The content R_k of the element when it is in the particular state (q_k , h_k) is defined by

$$R_k = \int_0^{q_k} h_k \, dq_k \qquad (5.1)$$

The dual quantity obtained by interchanging q_k and h_k is the co-content C_k of the element when it is in the particular state (h_k , q_k),

$$C_k = \int_0^{h_k} q_k \, dh_k \qquad (5.2)$$

The quantities defined above have the physical dimensions $[MLT^{-1}]$, i.e. power and may be interpreted as complementary forms of the power dissipation $q_k h_k$. From equations 5.1 and 5.2 one obtains:

$$R_k + C_k = \int_0^{q_k} h_k \, dq_k + \int_0^{h_k} q_k \, dh_k \qquad (5.3)$$

A transformation of variables and integration by parts of the second term in equation 5.3 produces

$$R_k + C_k = q_k h_k + \int_0^{q_k^{(1)}} h_k \, dq_k \tag{5.4}$$

where $q_k^{(1)} = q_k^{(0)}$.

From equation 5.4 the following relationships between contents, co-contents and power dissipation result:

For pipes or valves,

$$R_k + C_k = q_k \, h_k \tag{5.5}$$

i.e. content + co-content = power dissipation.

For pumps,

$$R_k + C_k = q_k \, h_k + A_k \tag{5.6}$$

i.e. content + co-content = power dissipation + constant. A_k is a negative constant for each pump, numerically equal to the total area limited by the characteristic curve and the co-ordinate axes in the quadrant $q \geq 0, h \leq 0$.

For ideal check valves,

$$R_k = 0 , \quad C_k = 0 , \quad q_k \, h_k = 0 \tag{5.7}$$

By extension, the content of a network when it is in the state $(\{q\}, \{h\})$ is defined to be the sum of the contents of it's elements, or,

$$R = \sum_{k \in K} R_k = \sum_{k \in K} \int_0^{q_k} h_k \, dq_k \tag{5.8}$$

where K is the set of element labels.

Similarly, the co-content of the network is defined by the expression

$$C = \sum_{k \in K} C_k = \sum_{k \in K} \int_0^{h_k} q_k \, dh_k \tag{5.9}$$

5.3   Optimization Model for the Solution of Analysis Problem

For the general formulation of the analysis problem the following assumptions are made:

1. The network is connected and closed.
2. Pipes, pumps and valves have unrestricted monotonic increasing charateristics represented by differential functions.

5.3.1 Non-linear programming formulation

Consider a network of nn nodes and ne elements consisting of element types 1, 2 and 5. The non-linear programming formulation of the problem becomes:

Determine q, h which satisfy the conditions:

$$
\begin{aligned}
& Nq = 0 \\
& Lh = 0 \\
& h_i = h_i^* & & i \in ne1 \\
& q_i = q_i^* & & i \in ne2 \\
& h_i = h_i\,(q_i) \ \ \text{or} \ \ q_i = q_i\,(h_i) & & i \in ne5
\end{aligned}
\tag{5.10}
$$

where:

q	:	ne × 1 vector of element flows
h	:	ne × 1 vector of element head losses
N	:	nn × ne node incidence matrix
L	:	nl × ne loop incidence matrix
ne1	:	number of element type 1
ne2	:	number of element type 2
ne5	:	number of element type 5
i	:	indices

5.3.2 State equations and state variables

Consider a spanning tree T_{1a} having chords made by elements type 2 and b. Nodal continuity equation, $Nq = 0$, can be expressed as,

$$
q_1 = L_{21}^T q_2^* + L_{b1}^T q_b
\tag{5.11}
$$

$$
q_a = L_{2a}^T q_2^* + L_{ba}^T q_b
\tag{5.12}
$$

The loop head loss equations, $Lh = 0$, can be written as,

$$
h_2 + L_{21} h_1^* + L_{2a} h_a = 0
\tag{5.13}
$$

$$h_b + L_{b1} h_1^* + L_{ba} h_a = 0 \qquad (5.14)$$

The above four equations represent the state equations for the problem in which the only independent variable is q_b.

5.3.3 Partial direct solution and partial iterative solution

From equations 5.11 and 5.12 it is obvious that once q_b is known, q_1 and q_a can be determined directly. And after knowing q_a and, thus, h_a, h_2 can be evaluated directly. Hence, the solution of the system can be divided in two parts;

1. Iterative solution for q_b
2.(i) Direct solution for q_1
    (ii) Direct solution for q_a (and thus h_a)
    (iii) Direct solution for h_2

This implies if elements type 1 and 2 are pressure-constrained and flow-constrained physical elements of the network, the evaluation of unknown design parameters for these element is done directly and these parameters are not the part of iterative solution.

5.3.4 Optimization model

Equations 5.10 are precisely the Kuhn-Tucker necessary conditions to be satisfied by the optimal solutions (q) and (h) of the pair of non-linear programming problems:

Primal minimization problem (in terms of flow variables)

Determine q which minimises

$$\phi = \phi(q) = \sum_{i=0}^{ne} \int_0^{q_i} h_i(q_i) \, dq_i = R \qquad (5.15)$$

subject to the constraint

$$Nq = 0 \qquad (5.16)$$

The quantity R is the total content of the network defined in section 5.2. In other words, problem can be stated as "Determine a distribution of flows satisfying Kirchhoff's node law which minimises the total content of the network".

In terms of flow variable q_b, the total content and, thus, the problem can be defined as,

Determine q_b, which minimises

$$\phi(q_b) = \sum_{i=0}^{neb} \int_0^{q_{b_i}} h_{b_i}(q_{b_i})\, dq_{b_i} + \sum_{i=0}^{nea} \int_0^{q_{a_i}} h_{a_i}(q_{a_i})\, dq_{a_i} + \sum_{i=0}^{ne1} h_{1_i}^{*T} q_{1_i} \tag{5.17}$$

subject to

$$q_1 = L_{21}^T q_2^* + L_{b1}^T q_b \tag{5.18}$$

$$q_a = L_{2a}^T q_2^* + L_{ba}^T q_b \tag{5.19}$$

where last term in equation 5.17 is the power input at elements where a condition related to type 1 elements prevail.

Dual minimization problem (in terms of head loss variables)

Determine h which minimises

$$\psi = \psi(h) = \sum_{i=0}^{ne} \int_0^{h_i} q_i(h_i)\, dh_i = C \tag{5.20}$$

subject to the constraint

$$Lh = 0 \tag{5.21}$$

The quantity C is the total co-content of the network. Problem can be defined as " Determine a distribution of head losses satisfying Kirchhoff's loop law, which minimises the total co-content of the network".

In terms of head variables, h_a, the total co-content and, thus, the problem can be defined as,

Determine h_a which minimises

$$\psi(h_a) = \sum_{i=0}^{neb} \int_0^{h_{b_i}} q_{b_i}(h_{b_i})\, dh_{b_i} + \sum_{i=0}^{nea} \int_0^{h_{a_i}} q_{a_i}(h_{a_i})\, dh_{a_i} + \sum_{i=0}^{ne2} q_{2_i}^{*T} h_{2_i} \tag{5.22}$$

subject to

$$h_2 + L_{21} h_1^* + L_{2a} h_a = 0 \tag{5.23}$$

$$h_b + L_{b1} h_1^* + L_{ba} h_a = 0 \tag{5.24}$$

The last term in equation 5.22 is the power input at elements where a type 2 elements prevail i.e. fixed supply or demand type elements.

It can be shown that under the assumptions stated at the beginning of the section, ϕ (q_b) and $\psi(h_a)$ are convex functions and also that the Kuhn-Tucker necessary conditions (5.10) are sufficient for the optimality of q^* in the primal and of h^* in the dual formulations (refer appendix B).

Rephrased, this means that any solution (q^*, h^*) of the analysis problem corresponds to an optimal solution q^* of the primal problem and to an optimal solution h^* of the dual problem and conversely. Moreover, solution is unique as the network obeys necessary and sufficient conditions as described in chapter 4.

5.4   Optimization Model for the Solution of Analysis-Design Problem

Analysis-Design problem consists of elements type 1, 2, 3, 4 and 5. The assumptions made for analysis problem, given in section 5.3, are also applicable to analysis-design problem.

5.4.1   Non-linear programming formulation

Consider a network of nn nodes and ne elements consisting of element types 1, 2, 3, 4 and 5. The non-linear programming formulation of the problem becomes:

Determine q, h which satisfy the conditions:

$$
\begin{aligned}
&Nq = 0 \\
&Lh = 0 \\
&h_i = h_i^* && i \in ne1 \\
&q_i = q_i^* && i \in ne2 \\
&h_l = h_l^* && i \in ne3 \\
&q_i = q_i^* && i \in ne3 \\
&h_i = h_i (q_i) \quad \text{or} \quad q_i = q_i (h_i) && i \in ne5
\end{aligned}
\tag{5.25}
$$

where ne3 is number of type 3 elements.

5.4.2   State equations and state variables

Consider a spanning tree T_{14a} having chords made by elements type 2, 3 and b. Nodal continuity equation, $Nq = 0$, can be expressed as,

$$
q_1 = L_{21}'^{T} q_2^* + L_{31}'^{T} q_3^* + L_{b1}'^{T} q_b
\tag{5.26}
$$

$$
q_4 = L_{24}'^{T} q_2^* + L_{34}'^{T} q_3^* + L_{b4}'^{T} q_b
\tag{5.27}
$$

$$q_a = L_{2a}^{\prime T} q_2^* + L_{3a}^{\prime T} q_3^* + L_{ba}^{\prime T} q_b \tag{5.28}$$

Consider spanning tree T_{13a} having chords made by elements type 2, 4 and b. The loop head loss equations, $Lh = 0$, can be written as,

$$h_2 + L_{21} h_1^* + L_{23} h_3^* + L_{2a} h_a = 0 \tag{5.29}$$

$$h_4 + L_{41} h_1^* + L_{43} h_3^* + L_{4a} h_a = 0 \tag{5.30}$$

$$h_b + L_{b1} h_1^* + L_{b3} h_3^* + L_{ba} h_a = 0 \tag{5.31}$$

The above six equations represent the state equations for the problem in which the only independent variable is q_b.

5.4.3   Partial direct solution and partial iterative solution

From equations 5.26, 5.27 and 5.28 it is obvious that once q_b is known, q_1, q_4 and q_a can be determined directly. And after knowing q_a and, thus, h_a, h_2 and h_4 can be evaluated directly. Hence, the solution of the system can be divided in two parts;

1.   Iterative solution for q_b
2.(a) Direct solution for q_1
   (b) Direct solution for q_4
   (c) Direct solution for q_a (and thus h_a)
   (d) Direct solution for h_2
   (e) Direct solution for h_4

This implies that the evaluation of unknown design parameters for element type 1, 2, and 4 are done directly and these parameters are not the part of iterative solution.

5.4.4   Optimization model

Equations 5.25 are precisely the Kuhn-Tucker necessary conditions to be satisfied by the optimal solutions (q) and (h) of the pair of non-linear programming problems:

Primal minimization problem (in terms of flow variables)

Considering spanning tree T_{13a} and in terms of flow variable q_b, the total content and, thus, the problem can be defined as,

Determine q_b which minimises

$$\phi(q_b) = \sum_{i=0}^{neb} \int_{0}^{q_{b_i}} h_{b_i}(q_{b_i}) \, dq_{b_i} + \sum_{i=0}^{nea} \int_{0}^{q_{a_i}} h_{a_i}(q_{a_i}) \, dq_{a_i} + \sum_{i=0}^{ne1} h_{1_i}^{*T} q_{1_i} + \sum_{i=0}^{ne3} h_{3_i}^{*T} q_{3_i}^{*} \quad (5.32)$$

subject to

$$q_1 = L_{21}^T q_2^* + L_{41}^T q_4 + L_{b1}^T q_b \quad (5.33)$$

$$q_3^* = L_{23}^T q_2^* + L_{43}^T q_4 + L_{b3}^T q_b \quad (5.34)$$

$$q_a = L_{2a}^T q_2^* + L_{4a}^T q_4 + L_{ba}^T q_b \quad (5.35)$$

where last two terms in equation 5.32 is the power input at elements where a condition related to type 1 and 3 elements prevail.

Dual minimization problem (in terms of head loss variables)

In terms of head loss variables, h_a, the total co-content and, thus, the problem can be defined as,

Determine h_a which minimises

$$\psi(h_a) = \sum_{i=0}^{neb} \int_{0}^{h_{b_i}} q_{b_i}(h_{b_i}) \, dh_{b_i} + \sum_{i=0}^{nea} \int_{0}^{h_{a_i}} q_{a_i}(h_{a_i}) \, dh_{a_i} + \sum_{i=0}^{ne2} q_{2_i}^{*T} h_{2_i} + \sum_{i=0}^{ne4} q_{4_i}^{T} h_{4_i}$$

$$(5.36)$$

subject to

$$h_2 + L_{21} h_1^* + L_{23} h_3^* + L_{2a} h_a = 0 \quad (5.37)$$

$$h_4 + L_{41} h_1^* + L_{43} h_3^* + L_{4a} h_a = 0 \quad (5.38)$$

$$h_b + L_{b1} h_1^* + L_{b3} h_3^* + L_{ba} h_a = 0 \qquad (5.39)$$

The last two terms in equation 5.36 is the power input at elements where a type 2 and 4 elements prevail.

It can be shown that under the assumptions stated at the beginning of the section, ϕ (q_b) and $\psi(h_a)$ are convex functions and also that the Kuhn-Tucker necessary conditions (5.25) are sufficient for the optimality of q^* in the primal and of h^* in the dual formulations (refer appendix B).

5.5   Method of solution

The choice of a method of solution is generally conditioned by the particular structure of the problem considered. The non-linear programming problems considered in sections 5.3 and 5.4 have the following particularities of importance to the implementation of a method of solution:

a. The objective functions ϕ and ψ are strictly convex.
b. The constraints, when considered explicitly, are linear.
c. First and second order information (gradients and hessians of the objective functions) can, in most cases, be obtained at a relatively low cost. This follows from the fact that most of the elements in the network have characteristics represented by simple explicit analytic relations between the two variables, flow and head loss.
d. The hessian matrices are sparse, i.e., contain only a relatively small number of non-zero entries, mostly in the case of problems with a large number of variables.

Among the many traditional methods of solution of non-linear programming problems available in the literature on optimization (Polak, 1971; Fletcher, 1987; Gue, 1971 among others), the particular class of descent methods has characteristics which are convenient for the application to problems with the features described above. Numerical surveys available in the literature show that descent methods perform efficiently in convex unconstrained minimization problems. However, it is also known how these methods can be extended to include linear constraints.

A canonical genetic algorithm (Holland, 1975: Goldberg, 1989) received lately a lot of attention and has been applied to find global optimum in a variety of problems. This algorithm is based on the idea of modelling the search process of natural evolution, although these models are crude simplifications of biological reality. Considering the properties of the objective functions of the two network analysis and analysis-design problems, genetic algorithm can well be applied to find the optimal solutions.

In this section, details of different methods available to solve non-linear programming problems and their relative importance are not given. For details, the above referred and other available literature may be referred.

5.6   Solution of Analysis and Analysis-Design Problems

Solution of analysis and analysis-design problems involves following steps:

1. Modelling of network component as elements type 1, 2, 3, 4 and 5. This depends on the location of specifications and objective of the problem.

2. Generation of two spanning trees, T_{13a} and T_{14a}. An algorithm for the generation of these two spanning trees is given in Appendix C.

3. Iterative solution of flows in elements type b.

4. Direct solution of flow and head loss in all other components.

5. Direct solution for head at nodes.

6. Direct solution for estimation of design parameters.

5.7   Applications of Analysis and Analysis-Design Problems

Analysis and analysis-design problems find applications in the variety of network design, operation and calibration problems. Some of the applications are as follows:

1. Determination of design parameters of network components for the specified outflows and/or heads at nodes. Design parameters may be diameter of pipes, water tank levels, pump power etc.

2. Determination of operating parameters of network components such as valve coefficients, pump speed etc. for the specified outflows and/or heads at nodes.

3. Determination of outflows or demand for the specified heads.

4. Determination of calibration parameters such as roughness coefficient from the measured data of outflow and head at different locations.

5. Standard analysis of network for the specified outflows reservoir levels.

The developed algorithms for the analysis and analysis-design problems provide direct estimation of required parameters.

Applications of the algorithms are shown using a pipe network as an example.

Consider a pipe network as shown in Fig. 5.1. Network has 30 pipes, 12 outflow points and one reservoir. At each outflow point there is a valve. The initial network component characteristics are given in Table 5.1 and initial steady state is given in Tables 5.2 and 5.3.

Table 5.1 Characteristics of network component shown in Fig. 5.1

Pipe No.	Length (m)	Diameter (m)	Pipe No.	Length (m)	Diameter (m)
1	300.	1.0	16	250.	0.65
2	250.	0.6	17	250.	0.6
3	450.	0.8	18	300.	0.55
4	150.	0.6	19	175.	0.55
5	150.	0.6	20	300.	0.6
6	160.	0.6	21	200.	0.6
7	140.	0.7	22	100.	0.7
8	140.	0.7	23	200.	0.6
9	170.	0.6	24	300.	0.6
10	170.	0.5	25	200.	0.5
11	300.	0.8	26	250.	0.5
12	300.	0.65	27	250.	0.5
13	300.	0.55	28	200.	0.55
14	200.	0.65	29	200.	0.65
15	250	0.55	30	200.	0.55
f = 0.016 for all pipes; Reservoir level = 150. M					

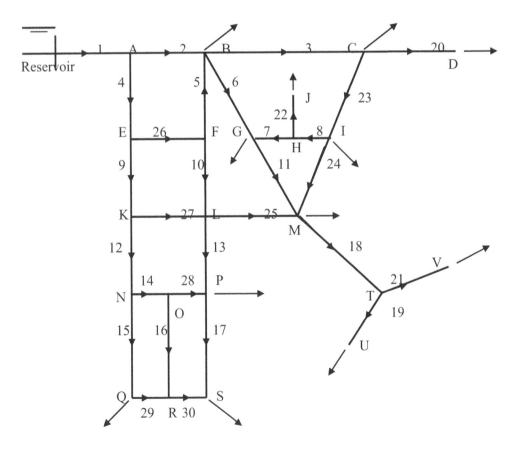

Fig. 5.1 A pipe network as an example

Table 5.2 Initial steady state of the network: flow in pipes and head at nodes

Pipe	q (m³/s)	Pipe	q (m³/s)	Node	H (m)	Node	H (m)
1	6.7	16	0.39	A	132.17	P	91.19
2	3.27	17	0.37	B	86.71	Q	81.69
3	3.52	18	1.1	C	64.23	R	81.68
4	3.43	19	0.5	D	62.39	S	90.6
5	1.54	20	0.6	E	102.12	T	73.72
6	0.79	21	0.2	F	92.75	U	72.57
7	1.0	22	0.2	G	85.01	V	72.50
8	1.2	23	2.42	H	86.14		
9	2.1	24	0.61	I	87.75		
10	0.2	25	0.1	J	86.10		
11	1.2	26	1.34	K	89.48		
12	1.3	27	0.77	L	93.03		
13	0.48	28	0.39	M	83.27		
14	0.78	29	0.03	N	83.54		
15	0.53	30	0.43	O	82.13		

Table 5.3 Initial outflows through valves

Valve at node	Outflow (m³/s)	Valve at node	Outflow (m³/s)	Valve at node	Outflow (m³/s)
B	0.5	I	0.6	Q	0.5
C	0.5	J	0.2	S	0.8
D	0.6	M	0.8	U	0.5
G	0.6	P	0.5	V	0.6

Following cases are considered to determine design, operation and calibration parameters.

Case A. Determination of diameters of pipes for the specified heads and outflows.

For this case, it is desired to determine the diameters of pipe no. 2, 3, 6, 8, 11, 13, 15, 17, 19, 20, 21 and 22 for the specified outflows and nodal heads as shown in Table 5.4. These 12 pipes are considered as type 4 elements and 12 nodes with specified heads and outflows are considered as type 3 nodes.

Table 5.4 Specifications for Case A: head and outflows at nodes

Node	Head (m)	Outflow (m³/s)	Node	Head (m)	Outflow (m³/s)
B	84.9	0.7	M	81.0	1.0
C	76.9	0.7	P	81.4	0.7
D	75.4	0.8	Q	80.6	0.7
G	82.3	0.8	S	81.0	0.9
I	82.4	0.8	U	67.5	0.6
J	82.3	0.4	V	67.3	0.7

Results, head loss and discharge, obtained for type 4 elements are used for the calculation of diameter of pipes using Darcy-Wiessbach equation. Results are listed in Table 5.5.

Table 5.5 Results for Case A: diameter of pipes

Pipe	Diameter (m)	Pipe	Diameter (m)	Pipe	Diameter (m)
2	0.8	11	0.95	19	0.85
3	0.95	13	0.85	20	0.70
6	0.85	15	0.75	21	0.80
8	0.95	17	0.80	22	0.80

Case B. Determination of valve coefficients for specified outflows and head at nodes.

For this case, it is desired to compute valve coefficients for specified outflows at nodes D, U and V and head at nodes as given in Table 5.6.

Table 5.6 Specifications for Case B : outflows and head at nodes

Node	Head (m)	Node	Head (m)	Node	Outflow (m^3/s)
A	125.7	M	58.5	D	0.9
C	27.6	P	79.4	U	0.8
H	63.8	R	57.6	V	0.9
I	66.5	T	35.7		
L	80.5				

In this case, node D, U and V are modelled as type 2 elements. Nodes C, I, M and P are modelled as type 1 elements. Nodes A, H, L, R and T are modelled as type 3 elements. Valves at nodes B, G, J, Q and S are modelled as type 4 elements.

Results of the analysis provide head loss and discharge in all the elements. Based on these results dimensionless valve coefficients are calculated. Table 5.7 lists head, outflow and valve coefficients at all outflow nodes. These valve coefficients are calculated considering the initial steady state as a reference.

Table 5.7 Results for Case B : head, outflows and valve coefficients

Node	Head (m)	Outflow (m³/s)	Valve Coefficient (τ)
B	64.31	0.62	1.44
C	79.38	0.57	1.74
D	23.50	0.90	2.44
G	61.87	0.63	1.23
I	66.54	0.61	1.17
J	63.81	0.22	1.28
M	58.53	0.71	1.06
P	79.38	0.53	1.14
Q	57.64	0.52	1.24
S	78.91	0.81	1.08
U	32.79	0.80	2.38
V	32.98	0.90	2.22

Case C. Determination of calibration parameters.

In this case, it is desired to evaluate roughness coefficients of pipes for model calibration. Measured values of outflows and head at nodes are the specifications. These specifications are given in Table 5.8. Objective of the problem is to compute Darcy-Wiessbach roughness coefficient for pipes 1, 2, 3, 9, 10, 12, 13, 15, 17, 18, 22 and 25.

Table 5.8 Specifications for Case C : outflows and head at nodes

Node	Head (m)	Node	Outflow (m³/s)
A	128.86	B	0.55
B	69.14	C	0.55
C	32.27	D	0.90
E	89.43	G	0.60
F	79.70	I	0.65
H	68.86	J	0.30
K	72.30	M	0.70
L	81.30	P	0.50
N	65.45	Q	0.55
O	63.62	S	0.80
R	63.09	U	0.80
T	41.26	V	0.90

In this case, nodes A, B, C, E, F, H, K, L, N, O, R and T are modelled as type 3 elements. Nodes D, G, I, J, M, P, Q, S, U and V are modelled as type 2 elements. Pipe elements for which roughness coefficient is to be evaluated are modelled as type 4 elements.

Results of the analysis provide head loss and discharge in pipes, which are modelled as type 4 elements. Based on these data, roughness coefficient of these pipes is calculated based on Darcy-Wiessbach equation. These coefficients are given in Table 5.9. Diameters of all pipes are taken as for initial condition.

Table 5.9 Results of Case C : roughness coefficient of pipes

Pipe	Roughness coefficient	Pipe	Roughness coefficient	Pipe	Roughness Coefficient
1	0.014	10	0.0165	17	0.013
2	0.015	12	0.015	18	0.016
3	0.0162	13	0.017	22	0.0155
9	0.017	15	0.018	25	0.013

These three cases are considered just to show the applications of analysis and analysis-design problems and the algorithm. A variety of other problems related to network design, operation and calibration can be solved using the developed algorithm.

5.8   Concluding Remarks

Simulation of steady state of hydraulic networks is the most basic requirement of network design and operation. Available codes such as EPANET, WATSYS etc. are based on analysis approach. Design of system components is carried out using trial and error procedures.

In this chapter, formulation of the steady state problem is carried out which is based on synthesis approach. Algorithm provides direct design of system parameters for the specified hydraulic behaviour of the system. Procedure developed is helpful in understanding the relationships between specifications and design variables.

Algorithm is based on two spanning trees T_{13a} and T_{14a}. This algorithm demarcates independent and dependent variables automatically once these two spanning trees are defined and provides a minimum set of independent variables. The other main advantage of the algorithm is the explicit calculation of the design parameters. Moreover, non-linear programming formulation of these problems in terms of minimisation of convex functions related to total content and co-content of the network provides a methodology, which guarantees convergence of the sought solution.

A variety of applications of these problems can be found in network design, operation and calibration for the cases where model is determined i.e. number of type 3 elements are equal to number of type 4 elements in the network. If number of type 4 elements are more than the number of type 3 elements, an optimisation problem has to be formulated based on minimisation of some objective such as cost of the network.

6 | Principles of Control of Pressure Surges

A wide range of unsteady flow problems in piping systems falls within the domain of rigid water column theory. The rigid water column approach to analyse transients with low frequencies in pipeline systems is a lumped approximation of the elastic column model (Watters, 1984). Though with the development of computer packages enabling precise calculations of transients in the systems using elastic water column approach, growing interest in the use of rigid model shows that rigid model is still helpful in the solution of some practical problems. Rigid model gives a simpler view of the way changes occur and have the advantage of immediately identifying some of the fundamental parameters of a given problem and explicitly demonstrating how the answers to the problems are influenced by variations in the parameters. Such a facility provides physical insight and significant implications not only in the educational process but also for many engineering design and analysis problems.

Problems of analysis and control of slow transients or pressure surges in complex pipe networks can be divided into two types. The first is the analysis problem in which system's hydraulic behaviour (i.e. the variation of flow and pressure with time) is determined for the specified valve operations. In analysis problems, no constraints in terms of head and/or flow specifications at the nodes or in the elements are imposed and, thus, the time integration of a system of ordinary differential equations of the first order which describes the network's behaviour during transient condition can be made in a straightforward manner by using a adequate numerical technique. The second is the synthesis problem. In this problem, valve operations are treated as variables and are designed to transfer the system from an initial steady state to a final steady state condition under specified transients. Procedure can be termed as transient design. Depending upon the number of specifications and control variables, control problems in networks are divided into two parts, determined problems and underdetermined problems.

In this chapter, control problems of determined nature are considered. Underdetermined problems are taken up in the next chapter. Algorithm for the analysis and control of pressure surges is developed for pipe networks. First a case of single reservoir-pipe-valve system is considered and then a general treatment of the problem for an arbitrary network is carried out. Among the advantages of the developed algorithm are automatic separation of dependent and independent variables, minimum set of independent variables and explicit design of valve operations.

6.1   Principles of Control of Pressure Surges : Transient Design

In pipeline systems, routine control of flow is often carried out by operating valves. Valves by introducing losses into a system control the rate of flow. If demand changes, it is necessary to define valve manoeuvres in order to optimise the operation, that is, to satisfy the demand and other specified objectives. Objectives to be fulfilled may be in terms of limitation on maximum/minimum head at specified locations or specifications of head/flow variations at specified locations in the network.

In existing pipeline systems, control of pressure surges is carried out by proper valve operations. However, problems related to design of new systems or expansion of existing systems may require design of some network components for the control of pressure surges besides design of valve operations. In general, problem of control of pressure surges can be stated as design of valve operations and system components to meet the demand while keeping the constraints or stated objectives satisfied.

Two approaches exist to solve the problem of control of pressure surges. First is analysis approach. This approach is based on trial and error procedure. In this approach, a suitable design of system component or valve operation is selected and system is analysed. If the resulting transients are not within stated limits or are not satisfying the constraints/objectives, a new design of system component/valve operation is selected for the system analysis. Procedure is repeated till hydraulic behaviour of the system is within satisfactory limits. Problem formulation using analysis approach is straightforward and a suitable numerical technique is used for the solution of a system of differential equations.

Second is synthesis approach in which parameters of a system component or valve operations are considered as variables and are designed directly for the specified transients at different locations in the system. Current practices for the design of valve operations or system components are generally based on analysis approach. Synthesis approach has not received much attention due to complexity involved in mathematical formulation of the problem.

Mathematically, the problem is equivalent to design of valve operations to transfer the system from an initial steady state to a final steady state under specified transients and can be termed as transient design.

Transient specifications can be of following nature:

i)   Specified head/head loss variations at network nodes or elements.
ii)   Specified flow/outflow variations in network elements or at nodes.
iii)   Specified both head/head loss and outflow/flow at nodes or elements.

Considering a closed network model, these three specifications constitute elements type 1, 2 and 3 respectively. Controlling valves whose operations are to be designed are type 4 elements. Valve can also be of element type 1 or 2 if head loss or flow variations through it are specified. All other network elements with known relationship between head loss and flow are type 5 elements. Hence, a network model consists of element type 1, 2, 3, 4 and 5.

Depending upon the number of specifications and control variables, two types of network control problems exists. First is the determined problem in which number of type 3 elements are equal to number of type 4 elements. Determined models should obey necessary and sufficient conditions for the existence and uniqueness of a solution as derived in chapter 4. Hence, two spanning trees, T_{13a} and T_{14a}, exist in determined models.

The second network control problem is underdetermined problem. In underdetermined models, number of type 4 elements is more than the number of type 3 elements. These models should also obey all necessary and sufficient conditions except second and seventh necessary conditions. However, in these models also two spanning trees, T_{13a} and T_{14a}, exist but the number of type a elements in spanning tree T_{13a} exceeds the number of type a elements in T_{14a} by the number type 4 elements exceeds type 3 elements. Thus some of the type a elements in spanning tree T_{13a} are type b elements in spanning tree T_{14a}.

Generally, valve operation strategies are to be designed for the specified time and the objective of the control problem is to transfer the network from an initial steady state to a final steady state within this specified time. However, in complex pipe networks it is always not possible to avoid residual transients after the valve operations cease. The capacity of the network to transfer itself from one steady state to another within specified time can be termed as network controllability. Controllability of a network is dependent not only on the network's characteristics and on topology but also on the number and location of valves. Therefore, it is necessary to derive conditions for the full controllability of a network, thus, guaranteeing the network to reach final steady state within specified time.

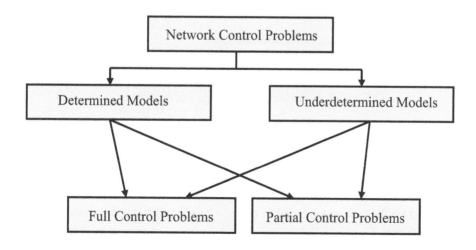

Fig. 6.1 Types of network control problems

Depending upon whether the number and location of valves satisfy the necessary condition for full controllability of a network, two types of control problems exist. The first is the full control problem in which valve topology and number guarantees full controllability of a network. The second is the partial control problem in which residual transients exist at the end of valve operations. In both problems, valve operations are designed for the specified transients or some objective criteria.

As shown in Fig. 6.1, network control problems are of four types: determined and underdetermined with full or partial controllability. In the next section, a single reservoir-pipe-valve system is considered. Thereafter, a general treatment is given to the determined problems in arbitrary networks.

6.2   Control of Pressure Surges in a Single Reservoir-Pipe-Valve System

Consider a single reservoir-pipe-valve system as shown in Fig. 6.2. The equation of unsteady flow in the pipe based on rigid water column theory is given by (Wylie, 1993)

$$h = H_A - H_B = m \frac{dq}{dt} + r q |q|$$

(6.1)

where

$$m = \frac{1}{Ag} \quad ; \quad r = \frac{fl}{2gDA^2}$$

(6.2)

m and r are inertial and resistance coefficients.

R: Reservoir
V: Valve

Fig. 6.2 A single reservoir-pipe-valve system

6.2.1   Frictionless case

If a uniform flow change from an initial steady state value, q_o , to a final steady state value, q_T , in time T is applied at valve end, there will be a sudden head change at time $t = 0$ and $t = T$ by a value equal to $m.(q_o - q_T)/T$ and a constant head during the entire time interval T.

For the case of full closure of valve the head variation at valve end is shown in Fig.6.3 by curve I. The resulting valve operations for this case are shown in Fig. 6.5 by curve I. Steep valve operations are obtained at time $t = 0$. However, for the case of partial closure, steep valve operations will also be obtained at time t = T. This is due to the fact that gradients of discharge at time t = 0 and t = T are not equal to zero.

To avoid sudden change in head, it is necessary that dq / dt should be zero at time t = 0 and t = T.

Any other discharge law can be prescribed at valve end (thus making the valve as element type 2) which will produce a change in head at valve end. Valve operations are designed after obtaining head variation at the valve from the analysis. However, in this case there is no control on head change at valve end, which depends on the way discharge is varied.

Consider the case of head variation at valve end as shown by curve II in Fig. 6.3 for the frictionless valve closure case. In this case, total time of operation, T, is divided into three parts. During time interval t_1 and t_2, head variation is linear and during the time interval t_3 head remains constant. The boundary conditions for discharge through the pipe are

At time t = 0,

$$q(0) = q_0 \; ; \quad \frac{dq}{dt}\bigg|_{t=0} = 0 \tag{6.3}$$

and at time t = T,

$$q(T) = q_T \; ; \quad \frac{dq}{dt}\bigg|_{t=T} = 0 \tag{6.4}$$

The discharge functions satisfying the above boundary conditions will become

$$q = \frac{ct^2}{2t_1} + q_o \qquad\qquad \text{for} \quad 0 \le t \le t_1 \tag{6.5}$$

$$q = c\left(t - \frac{t_1}{2}\right) + q_o \qquad\qquad \text{for} \quad t_1 \le t \le t_1 + t_3 \tag{6.6}$$

$$q = \frac{c}{2t_2}(2Tt - t^2 - T^2) + q_T \quad \text{for} \quad t_1 + t_3 \le t \le T \tag{6.7}$$

where

$$c = \frac{2(q_o - q_T)}{(t_1 + t_2 - 2T)} \tag{6.8}$$

The maximum head change at valve end, ΔH, is given by

$$\Delta H = \frac{2m(q_T - q_o)}{(t_1 + t_2 - 2T)} \tag{6.9}$$

For the case of $t_1 = t_2 = T/4$, the discharge function at valve is shown by curve II in Fig. 6.4. The valve operations for this case are shown by curve II in Fig. 6.5. For the purpose of calculations, m = 1 has been taken.

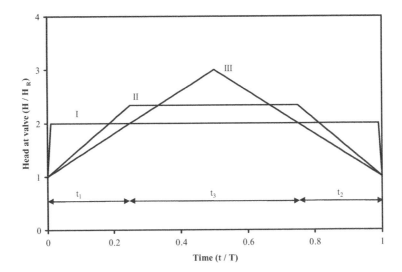

Fig. 6.3 Head variations at valve end due to valve closure for frictionless case

For the case when $t_1 = t_2 = 0$, $\Delta H = m.(q_o - q_T)/T$ which is equivalent to linear variation of discharge function. If $t_1 = t_2 = T/2$ and $t_3 = 0$, $\Delta H = 2m.(q_o - q_T)/T$ which is two times the head rise as compared to linear discharge variation case. For the case when $t_3 = 0$ and $t_1 = t_2 = T/2$, head and discharge variations at valve are given by curve III in Figs. 6.3 and 6.4 respectively. The resulting valve operations are shown by curve III of Fig. 6.5.

For a given ΔH and T, time t_1 and t_2 can be evaluated from equation 6.9 after fixing a ratio t_1/t_2 and then discharge functions can be calculated from equations 6.6-6.8.

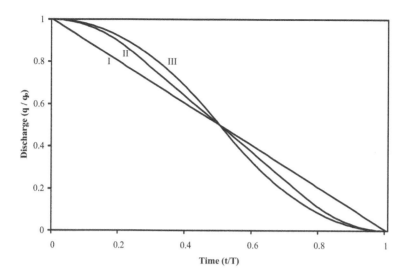

Fig. 6.4 Discharge variations at valve end due to valve closure for three cases of head variations (frictionless case)

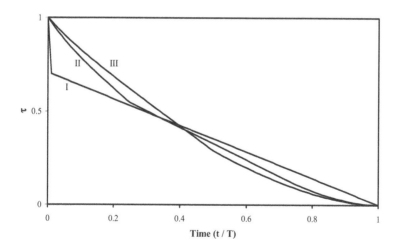

Fig. 6.5 Dimensionless valve coefficient for three cases of head variations (frictionless case)

Knowing discharge and head variations at valve end, valve coefficient can be calculated using

$$\frac{q_v}{q_{v_o}} = \tau \sqrt{\frac{h_v}{h_{v_o}}} \tag{6.10}$$

τ is the dimensionless valve coefficient and is equal to $(Cd.A_v)/(Cd_o.A_{vo})$. Subscript o stands for initial steady state condition.

6.2.2   Linearised frictional case

Dividing the total time T in small time steps Δt and considering a small change in discharge in each time step, the non-linear frictional term in eq. 6.1 can be linearised to $r'q(t)$ where $r' = r.q(t-\Delta t)$. Hence, eq. 6.1 becomes

$$h = H_A - H_B = m \frac{dq}{dt} + r'q \tag{6.11}$$

Considering a linear uniform discharge variation at valve end as shown by curve I in Fig, 6.7, the analysis provides a head variation at valve end as shown by curve I in Fig. 6.6. There is a sudden change in head at time $t = 0$ and $t = T$ as dq/dt is not equal to zero at time $t = 0$ and $t = T$. The resulting valve operations are shown by curve I in Fig. 6.8.

Consider the case of valve closure with a trapezoidal head variation at valve end as shown by curve II in Fig. 6.6. The discharge variation that satisfies boundary conditions as stated by eqs. 6.3 and 6.4 for this head variation will become

$$q = c_1 e^{-\alpha t} + c_2 t + c_6 \quad \text{for} \quad 0 \le t \le t_1 \tag{6.12}$$

$$q = c_3 e^{-\alpha t} + c_7 \quad \text{for} \quad t_1 \le t \le t_1 + t_3 \tag{6.13}$$

$$q = c_4 e^{-\alpha t} + c_5 t + c_8 \quad \text{for} \quad t_1 + t_3 \le t \le T \tag{6.14}$$

where $\alpha = r'/m$ and coefficients $c_1, c_2, c_3, c_4, c_5, c_6, c_7$ and c_8 are given by

$$c_1 = \frac{(q_T - q_o)}{\alpha \left[\dfrac{e^{-\alpha(t_1+t_3)} - e^{-\alpha t_3}}{e^{-\alpha(t_1+t_3)} - e^{-\alpha T}} e^{-\alpha T} . t_2 + t_1 \right]} \tag{6.15}$$

$$c_2 = c_1 \alpha \tag{6.16}$$

$$c_3 = c_1 (1 - e^{\alpha t_1}) \tag{6.17}$$

$$c_4 = c_1 \frac{e^{-\alpha(t_1+t_3)} - e^{-\alpha t_3}}{e^{-\alpha(t_1+t_3)} - e^{-\alpha T}} \tag{6.18}$$

$$c_5 = c_4 \alpha e^{-\alpha T} \tag{6.19}$$

$$c_6 = q_o - c_1 \tag{6.20}$$

$$c_7 = c_1 \alpha t_1 + q_0 \qquad (6.21)$$

$$c_8 = q_T - (1 + \alpha T) c_4 e^{-\alpha T} \qquad (6.22)$$

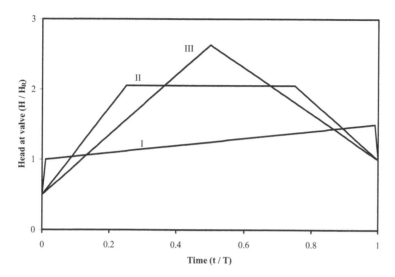

Fig. 6.6 Head variations at valve end due to valve closure for linearised friction case

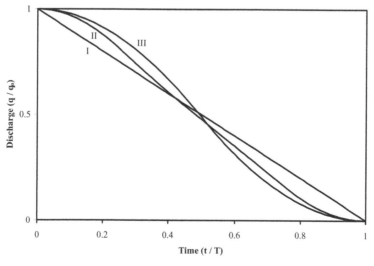

Fig. 6.7 Discharge *variations at valve end due to valve closure for* three cases of head variations (linearised *friction case)*

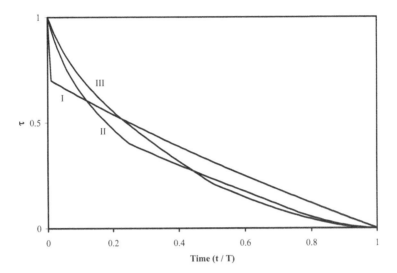

Fig. 6.8 Dimensionless valve coefficient for three cases of head variations (linearised friction case)

Using eq. 6.11, the maximum head rise at valve end above the initial steady state head is given by

$$\Delta H = \frac{-(q_T - q_0) r' t_1}{\left[\dfrac{e^{-\alpha(t_1 + t_3)} - e^{-\alpha t_3}}{e^{-\alpha(t_1 + t_3)} - e^{-\alpha T}} e^{-\alpha T} . t_2 + t_1 \right]} \tag{6.23}$$

For the specified maximum head change ΔH at valve end, known initial and final steady state discharge and time T, time t_1, t_2 and t_3 can be evaluated after fixing two time intervals out of three. Discharge functions can be calculated using eqs. 6.12 -6.14 and then valve operation rules using eq. 6.10.

For the case of full closure and $t_1 = t_2 = T/4$, the discharge function and resulting valve operations are shown by curve II in Figs. 6.7 and 6.8 respectively. For the case of $t_1 = t_2 = T/2$ i.e. for the head variation at valve as shown by curve III in Fig. 6.6, the discharge variation at valve and resulting valve operations are given by curve III in Figs. 6.7 and 6.8 respectively.

Specification of head variation at valve end makes valve as element type 1 and for linearised frictional case flow through pipe can be evaluated by analytical expression as derived above. This analysis provides no residual transients.

For the case of non-linear friction, eq. 6.1 is to be integrated by using an adequate numerical technique. In this case, residual transients will exist for any specification of head variation at valve end, as boundary conditions given by eq. 6.4 can not be satisfied. However, for non-linear case, if valve is modelled as type 2 element i.e. a discharge variation is specified through the valve which satisfies boundary conditions given by eqs. 6.3 and 6.4, no residual transients exist. In this case, however, there is no control on head variation at valve end. Head variation depends on the specified discharge function.

6.3   Control of Pressure Surges in Networks

As described in section 6.1, control of pressure surges on the principles of transient design involves design of valve operations for the specified transients in network elements. Consider a closed network consisting of elements type 1, 2, 3, 4 and 5 in which number of type 3 elements are equal to the number of type 4 elements.

6.3.1   Mathematical formulation for determined model

Consider two spanning trees, T_{14a} and T_{13a} in the closed network. The continuity equations based on spanning tree T_{14a} can be written as

$$q_1 = L'_{21} q_2^* + L'_{31} q_3^* + L'_{b1} q_b \tag{6.24}$$

$$q_4 = L'_{24} q_2^* + L'_{34} q_3^* + L'_{b4} q_b \tag{6.25}$$

$$q_a = L'_{2a} q_2^* + L'_{3a} q_3^* + L'_{ba} q_b \tag{6.26}$$

The loop head loss equations based on spanning tree T_{13a} are

$$h_2 + L_{21} h_1^* + L_{23} h_3^* + L_{2a} h_a = 0 \tag{6.27}$$

$$h_4 + L_{41} h_1^* + L_{43} h_3^* + L_{4a} h_a = 0 \tag{6.28}$$

$$h_b + L_{b1} h_1^* + L_{b3} h_3^* + L_{ba} h_a = 0 \tag{6.29}$$

The equations of motion for element type a and b are

$$h_a = m_a \frac{dq_a}{dt} + r_a q_a |q_a| \tag{6.30}$$

$$h_b = m_b \frac{dq_b}{dt} + r_b q_b |q_b| \tag{6.31}$$

m_a and r_a are inertia and coefficient of resistance matrices for type a elements. These are square and diagonal matrices of the order of ne_a by ne_a. Similarly m_b and r_b are inertia and coefficient of resistance matrices for type b elements and are square, diagonal matrices of the order of ne_b by ne_b. These matrices are given by,

$$m_{a_i} = \frac{l_{a_i}}{g\,A_{a_i}} \;;\; m_{b_j} = \frac{l_{b_j}}{g\,A_{b_j}} \tag{6.32}$$

$$r_{a_i} = \frac{f_{a_i} l_{a_i}}{2g\,D_{a_i} A_{a_i}^2} \;;\; r_{a_j} = \frac{f_{a_j} l_{a_j}}{2g\,D_{a_j} A_{a_j}^2} \tag{6.33}$$

where i (=1,2,….ne_a) and j (=1,2,….ne_b) are indices. Inserting the value of q_a from eq.6.26 into eq. 6.30 and then using the resulting equation for h_a along with the value of h_b from eq. 6.31 in the eq. 6.29, after rearranging the terms, yields

$$\frac{dq_b}{dt} = -K^{-1}\left[F_2 + L_{ba}m_a\frac{dF_1}{dt} + r_b q_b |q_b| + L_{ba}r_a\left(F_1 + L'^T_{ba}q_b\right)\left|\left(F_1 + L'^T_{ba}q_b\right)\right| \right] \tag{6.34}$$

where matrices K, F_1 and F_2 are given by

$$K = m_b + L_{ba}m_a L'^T_{ba} \tag{6.35}$$

$$F_1 = L'^T_{2a}q_2^* + L'^T_{3a}q_3^* \tag{6.36}$$

$$F_2 = L_{b1}h_1^* + L_{b3}h_3^* \tag{6.37}$$

dF_1/dt is the time derivative of F_1 matrix. If the quantities, q_2^* and q_3^*, are kept constant during the transient period, dF_1/dt vanishes. Equation 6.34 is the ordinary differential equation in which q_b is the only variable and the number of equations are equal to the number of type b elements i.e. ne_b.

In addition, equation of motion for water level variations in the surge tank is given by

$$q_s = A_s \frac{dH_s}{dt} \tag{6.38}$$

A_s is the area of surge tank. q_s and H_s are the discharge and water level in the surge tank or the head at node having surge tank. Here, the nomenclature for head at nodes for those type 5 nodes that have surge tanks has been changed from H_5 to H_s for better clarity of variables. Similarly, the nomenclature for discharge in those type a or b pseudo-elements that connect surge tank nodes with the reference node has been changed to q_s.

In eq. 6.38, H_s is the additional variable and the number of equations are equal to number of surge tanks in the network. In fact, variable H_s can be eliminated while deriving the final set of ordinary differential equations, which contains the only independent variable q_b. However, in this case, the set of ordinary differential equations will be of second order and the number of equations are equal to number of type b elements i.e. ne_b.

If the additional variable H_s is kept, the system of ordinary differential equations are of first order but the number of equations are increased by the number of surge tanks in the network.

6.3.2   Solution algorithm : partial numerical solution and partial direct solution

From eqs. 6.34 and 6.38, q_b and H_s are evaluated by numerical integration by using an suitable numerical technique such as Runge Kutta fourth order method. Once q_b and H_s are known, q_1, q_4 and q_a are calculated directly from eqs. 6.24, 6.25 and 6.26 respectively. H_a is calculated from eq. 6.30 and then h_2, h_4 and h_b are calculated directly from eqs. 6.27, 6.28 and 6.29 respectively.

The general solution algorithm for the control of pressure surges in networks by valve operations consists of following steps:

i)   Construct a network model consisting of elements type 1, 2, 3, 4 and 5 after imposing transient specifications in terms of head loss and/or discharge variations.

ii)   Check the solvability of the model. Network model should satisfy necessary and sufficient conditions for the existence and uniqueness of a solution as stated in chapter 4.

iii)   Define two spanning trees, T_{13a} and T_{14a}. This automatically defines independent variables q_b and H_s.

iv)   Knowing initial values for the independent variables and transient specifications, eqs. 6.34 and 6.38 are solved using a numerical technique. Numerical integration provides time histories of independent variable, q_b and H_s.

v)   Knowing q_b and H_s, all other variables are calculated directly by using eqs. 6.24-6.30.

vi)   After knowing values of q_4, h_4, h_2, q_1, valve operation rules are calculated using eq. 6.10.

6.3.3   Advantages of the algorithm

Following are the advantages of the developed algorithm for the control of pressure surges in networks by valve operations:

i)   Algorithm is developed for an arbitrary network and incorporates possible boundary conditions in terms of transient specifications of head loss and discharge variations.

ii)   Once two spanning trees, T_{13a} and T_{14a}, are defined, automatic separation of independent and dependent variables occurs.

iii)   It provides a minimum set of independent variables and, thus, minimum set of ordinary differential equations.

iv)   Numerical integration of ordinary differential equations containing only the independent variables is carried out. All other variables are calculated directly.

v)   Valve coefficients are not the part of independent variables and valve operation rules are calculated explicitly after evaluation of independent variables through numerical integration of ordinary differential equations.

6.4   Network Controllability : Partial and Full Control Problems

One basic question that arises in the control of pressure surges in a network is; Is it possible to transfer a system from any initial steady state to any other final steady state in a specified time without any residual transients by some proper valve operations?. The answer to this question depends on the controllability of a network.

Controllability of a network can be defined as its capacity to transfer itself from an initial steady state to a final steady state within specified time. If a system can be transferred to a final steady state condition, it is fully controllable and if network does not reach a final steady state condition at the end of valve operations and residual transients exist, it is called partially controllable.

From network state equations, it is clear that if all the state variables i.e. q_b can be controlled to reach a desired value in a specified time by some valve operations τ, network becomes fully controllable. Moreover, to avoid any residual transients in the system all dq_b/dt should be zero at the end of specified time of valve operations. Mathematically, these conditions can be stated as $q_b(T) = q_{bT}$ and $dq_b(T)/dt = 0$.

Valve operations satisfying these conditions guarantee full controllability of a network or in other words valve operations obtained from eq. 6.10 for those q_b which obey these conditions guarantee full control of pressure surges. Obviously it is possible only if all chords i.e., type b elements are valve elements. Hence, it is necessary for full controllability of a network that number of valves should be equal to the number of chords in the system and their locations in the network should be such that a spanning tress exists if all valve elements are removed.

Network control problems can be divided in two parts depending on whether the network is fully controllable or partially controllable. These problems can be termed as partial control problem and full control problem.

Obviously, if the network has more valves than required from full controllability condition, some of the valve's operation rules will be dependent on the operation rules of those valves that are necessary for full controllability.

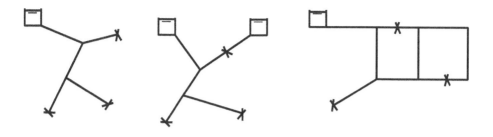

Fig. 6.9 Examples of fully controllable networks

Fig. 6.9 shows few examples of fully controllable networks. If any valve in a network is removed, network becomes partially controllable.

6.5 Applications of Determined Models

Determined models are useful in many different applications of network analysis and control of pressure surges. Problems of design of valve operations for transferring the network from an initial steady state to a final steady state with the specified transients in terms of head loss and/or discharge variations in some parts of the network are of immense practical use. Practical applications are found in the fields of transmission and distribution networks, industrial piping systems, hydropower systems, pumping stations etc. Design of valve operations for changing the outflow at different locations in the network in the specified trajectories, for changing the outflows with the specified head loss in the elements or head at nodes are few examples of such applications.

Application of partial control determined problem

As an example, consider a network as shown in Fig. 6.10. Network consists of one reservoir, one surge tank and four control valves. Characteristics of the network are given in Table 6.1. It is desired to transfer the network from an initial steady state to a final steady state with the specified transients at node B and D i.e. with specified outflow and head variations at these nodes.

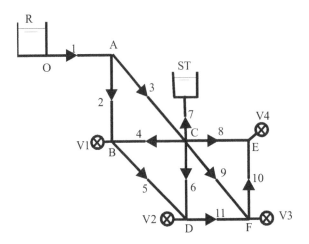

V : valve
R : reservoir
ST : surge tank

Number along the edges
shows pipe number

Fig. 6.10 A pipe network as an example

Table 6.1. Characteristics for Pipe Network in Fig. 6.4

Pipe	1	2	3	4	5	6	7	8	9	10	11
l (km)	1.0	1.0	1.5	1.0	1.0	1.5	.05	1.0	1.5	1.0	1.0
D (m)	2.8	0.95	2.45	1.2	1.5	1.3	1.0	0.85	1.0	1.3	1.25
f = 0.016 for all pipes ; Area of surge tank = 40 m²											

Table 6.2. Initial & Final Steady-States : Discharges, Heads and Valve Openings

Pipe	q (m³/s)		Node	Head (m)		Valve	q (m³/s)	
	Initial	Final		Initial	Final		Initial	Final
1	25.0	8.14	O	50.0	50.0	V1	5.0	1.23
2	4.03	1.27	A	45.19	49.49	V2	10.0	3.03
3	20.97	6.87	B	17.41	46.75	V3	4.0	1.99
4	5.80	1.78	C	35.30	48.43	V4	6.0	1.89
5	4.83	1.81	D	11.31	45.89	Valve opening (τ)		
6	8.20	2.67	E	4.55	44.76			
7	0.0	0.0	F	7.32	44.97	V1	1.0	0.15
8	3.21	1.11				V2	1.0	0.15
9	3.75	1.32				V3	1.0	0.20
10	2.79	0.78				V4	1.0	0.10
11	3.03	1.45						

Valves 1 and 3 are taken as type 3 elements in which a transient behaviour in terms of
linear increase of head loss from initial steady state value to final steady state in 30 in

30 seconds and a linear valve closure, $\tau = 1$ to 0.15 for valve 1 and $\tau = 1$ to 0.2 for valve 3 in 30 seconds are specified. Initial and final steady states of the network are given in Table 6.2.

Valves 2 and 4 are modelled as type 4 elements. All other elements except reservoir of the network are type 5 elements.

The resulting set of ordinary differential equations is numerically integrated using Runge Kutta fourth order method. A time step of 1 second is selected. Analysis is carried out for 300 seconds.

The resulting valve operation rules for valves 2 and 4 are shown in Fig. 6.11. Figs. 6.12 and 6.13 show discharge variations through valves and few pipes and nodal head variations respectively.

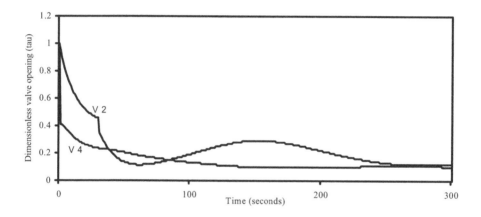

Fig. 6.11 Valve operation rules

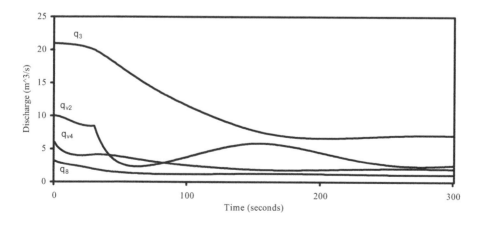

Fig. 6.12 Discharge variations in pipes and valves

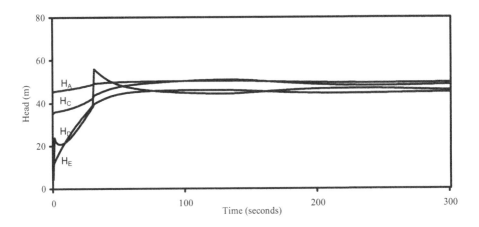

Fig. 6.13 Nodal head variations

Application of full control determined problem

Consider network shown in Fig. 6.14 as an example. Network has 24 pipes, 2 reservoirs and 12 valves. Network elements data is given in Table 6.3. It is desired to change the flow in the network from an initial steady state to a final steady state, as given in Tables 6.3, 6.4 and 6.5, in 60 seconds with following specifications:

1. Flow through valves 2, 5, 6, 7, 10 and 11 is changed smoothly based on the following equation:

$$q = q_0 - \frac{3(q_0 - q_T)}{T^2}t^2 + \frac{2(q_0 - q_T)}{T^3}t^3 \qquad (6.39)$$

These elements are, thus, modelled as type 2 elements.

2. Head variation at node B is given as shown in Fig. 6. 18 with a maximum head rise limited to 62 m. Hence, node B is modelled as type 3 node.

3. Pipe elements 5, 7, 15 and 16 are considered as type 3 elements. Discharge variation through these pipes is changed smoothly as given in Fig. 6.15. These pipe elements are also modelled as type 3 elements.

Valves 1, 3, 4, 8 and 9 are modelled as type 4 elements. All other pipe elements are modelled as type 5 elements.

Network has control valves in its every loop. Hence, the system can be transferred from initial steady state to final steady state in the given time without any residual transients.

Analysis is carried out using the developed algorithm. Valve operation rules are shown in Figs. 6.16 and 6.17. Nodal head variations are shown in Figs. 6.18 and 6.19.

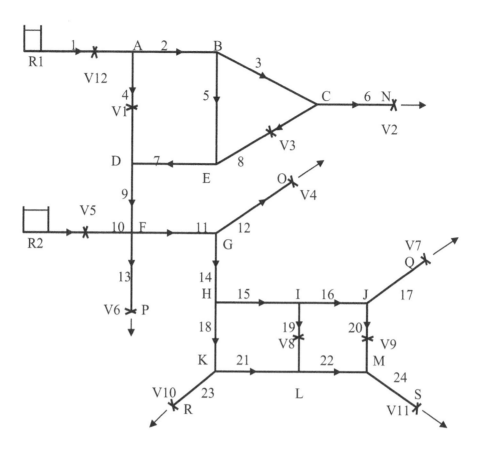

Fig. 6.14 A pipe network as an example

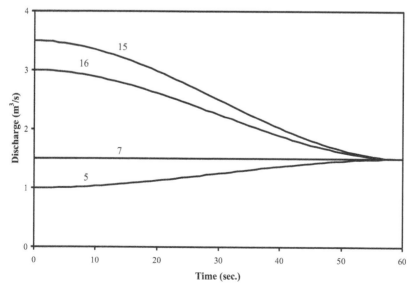

Fig. 6.15 Discharge variations in pipes as specifications

Table 6.3 Network pipe data and initial and final flows

Pipe No.	Length (m)	Diameter (m)	Initial flow (m^3/s)	Final flow (m^3/s)
1	100	1.0	4.0	3.0
2	100	0.6	2.5	1.5
3	150	0.8	1.5	0.
4	200	0.6	1.5	1.5
5	200	0.5	1.0	1.5
6	100	0.5	1.0	0
7	100	0.6	1.5	1.5
8	150	0.6	0.5	0.
9	150	0.8	3.0	3.0
10	100	1.0	5.5	4.5
11	150	1.5	6.5	6.5
12	100	0.6	1.0	2.0
13	150	0.6	2.0	1.0
14	150	1.5	5.5	4.5
15	150	1.0	3.5	1.5
16	150	1.0	3.0	1.5
17	100	0.55	2.5	1.5
18	100	0.6	2.0	3.0
19	100	0.6	0.5	0.
20	100	0.6	0.5	0.
21	150	0.6	1.0	1.0
22	150	0.6	2.0	1.0
23	150	0.6	1.0	2.0
24	100	0.6	2.0	1.0
f = .016 for all pipes; Level of reservoirs R1 = 70 m and R2 45 m				

Table 6.4 Initial and final valve coefficients

Valve No.	Initial opening (τ)	Final opening (τ)	Valve No.	Initial opening (τ)	Final opening (τ)
1	1.0	0.892	7	1.0	0.507
2	1.0	0.	8	1.0	0.
3	1.0	0.	9	1.0	0.
4	1.0	2.525	10	1.0	2.978
5	1.0	0.299	11	1.0	0.711
6	1.0	0.502	12	1.0	0.52

Table 6.5 Initial and final heads at nodes

Node	Initial head (m)	Final head (m)	Node	Initial head (m)	Final head (m)
A	67.89	63.52	K	30.8	14.05
B	57.25	59.98	L	28.24	11.5
C	55.89	60.35	M	22.5	9.16
D	44.95	36.67	N	51.65	60.77
E	48.78	40.5	O	36.69	23.02
F	39.5	31.22	P	29.29	29.11
G	38.39	30.12	Q	16.95	23.75
H	37.6	29.66	R	30.77	13.88
I	35.17	29.53	S	15.69	7.75
J	33.39	29.32			

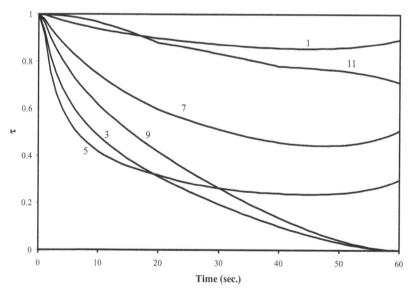

Fig. 6.16 Valve operation rules for valves 1, 3, 5, 7, 9 and 11

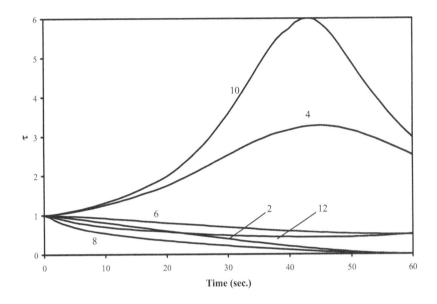

Fig. 6.17 Valve operation rules for valves 2, 4, 6, 8, 10 and 12

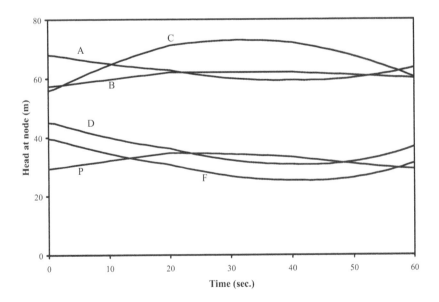

Fig. 6.18 Head variations at nodes

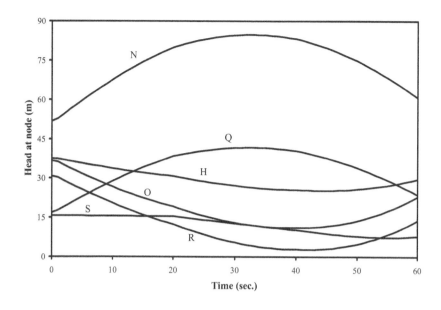

Fig. 6.19 Head variations at nodes

6.6   Concluding Remarks

In networks, control of flows is generally carried out through valve operations. Operational flow control problems involve determination of valve operation rules to transfer a network from an initial steady state to a final steady state. In many practical problems, it is desired to have a specified variation of outflows/discharges and/or head/head loss in some parts of the network during the transient period. In these problems, conventional procedures are based on analysis approach i.e. use of trial and error methods for the design of valve operations. Development of procedures for the direct design of valve operations for the specified transients i.e. synthesis approach is essential for avoiding tedious trial and error procedures.

Design of valve operations for the specified transients can be termed as transient synthesis or transient design. Depending upon whether the transient specifications are equal or less than the control variables, two types of control problems exist, determined problem and underdetermined problem. This chapter deals with the determined problems.

In complex networks, the mathematical formulation of such problems in a comprehensive and meaningful form is often neither immediately apparent nor straightforward. The problem of solvability of resulting models is dependent on the network flow distribution and on the manner in which the decision parameters and corresponding boundary specifications are topologically allocated. In this chapter, problem has been mathematically formulated for an arbitrary network consisting of possible transient specifications or boundary conditions. Determined models should obey necessary and sufficient conditions for the existence and uniqueness of a solution as derived in chapter 4.

The developed algorithm for the solution of determined models is more efficient as it provides a minimum set of ordinary differential equations with automatic separation of dependent and independent variables. Moreover, the valve operation rules are calculated explicitly.

Transfer of network from an initial steady state to a final steady state within specified time with no residual transients depends on the network controllability. For full controllability i.e. the case when no residual transients exist at the end of valve operations, each loop of the network should possess a control valve. If the network doses not possess valves in each loop, residual transients will persist.

From the example of a network, it is clear that in partial control problems though the transient specifications at desired locations are met while transferring the network from initial to final steady state, residual transients exist in the network. In such case, valve operates for a long time till the network attains final steady state. Such long operation of valves is generally undesirable. However, operation of valves can further be designed for the specified time. This problem has been discussed in chapter 8.

7 | Optimal Control of Pressure Surges by Valve Operations

Optimal control of pressure surges in pipe networks by valve operations involves transfer of the system from an initial steady-state to a final steady-state in a specified time without any residual transients at the end of valve operations with controlled pressure variations. Optimal control problems are full control problems in which network possess control valves in every loop. Optimal control problems of determined nature are solved directly as described in previous chapter. However, underdetermined problems require optimization of an objective function related to network.

In this chapter, a methodology based on optimisation techniques has been developed for the optimal control of pressure surges in networks containing valves in every loop. An objective function, which minimises the pressure change in the system and transfers the system to a final steady state smoothly is developed and optimised using principles of calculus of variations to get optimal valve operations.

7.1   Principles of Optimal Control

In general, a system is engineered with respect to specific performance requirements. Based on these specific performance requirements, the control problem is mathematically formulated and the optimum solution to the problem is sought.

In pipeline system, one of the main concerns is the maximum rise or drop in pressure during valve adjustments. Moreover, smooth transfer of a system from initial steady state to a final steady-state is desirable to avoid development of rapid transients. Consider a single reservoir– pipe–valve system shown in Fig. 6.2 as discussed in previous chapter. For the frictionless case, if discharge is varied linearly at valve end for a full closure case, there will be a sudden rise and drop of pressure equal to m.dq/dt at the valve at the beginning and end of valve operations as shown by curve I in Fig. 6.3. m.dq/dt is the minimum pressure rise/drop. If a maximum head rise at valve end, H_{msx},is given, discharge through valves can be obtained which give pressure variation shown by curve II in Fig. 6.3. In these cases, there is no control at rate of variation of head at the beginning and end of valve operation. For frictional case also, the same principle applies if the formulation is carried out with linearised friction.

In a network system, however, the problem is more complex. It is not always possible to guarantee head variation as shown by curve II of Fig. 6.3 at each node even if formulation is carried out with linearised friction. This head variation is possible for the cases when either all the nodes are outflow nodes having valves located there or all the pipes in the network have the same dimensionless coefficient rT/m. Generally these conditions are difficult to satisfy. Moreover limitation on head rise imposes limitation on rate of change of head and, thus, flow rate, Smooth change of flows is the main requirement if the system is being analysed by rigid water column theory, As pressure rise is the main functional requirements, design of valve operation should be made which minimises pressure rise as well as guarantee smooth flow changes. Such valve operations are called optimal valve operations, which controls pressure surges optimally.

In general, problem of optimal control of pressure surges in networks can be defined as the design of valve operations for transferring the network from an initial steady state to a final steady state within specified time such that the following requirements are satisfied:

i)   Minimum pressure change in the system or limitations on the maximum/minimum head change at specified locations in the network.
ii)   Smooth change of flows and heads in the system or limitations on the rate of change of heads or flows.

Shimada (1992) presented a methodology for the control of pressure surges by using optimal control theory. However, conditions for full controllability of a network were not presented and the methodology doesn't guarantee transfer of the system to a final steady state at the end of valve operations. Moreover, consideration for minimum pressure rise was not taken into account while deriving the operation strategy.

In this chapter, first state equations and state variables are derived which describe the dynamic behaviour of a system and then conditions for full controllability of a network are derived. An optimisation model is developed for the design of valve operations by minimising an objective function, which guarantees minimum pressure variation and a smooth transfer of system from a initial steady state to a final steady state. Procedure can be well termed as 'valve stroking for pressure surges'. Application of the present methodology is shown by examples of pipe network systems.

7.2   Mathematical Formulation : State Equations and State Variables

Consider a network consisting of valves in its each loop. All the valves are type 4 elements. If a valve is installed in the middle of a pipe, element can be divided into three parts: two pipe elements and one valve element. Three types of elements exist in an extended network. Elements joining reservoir nodes with the reference node are called type 1 elements. A spanning tree of the closed network consists of all type 1 elements and all pipe elements, which are type a elements.

By grouping fixed head nodes and variable head nodes separately and, similarly, all type 1, a and 4 elements, node incidence matrix for the network can be written as

$$N = \begin{bmatrix} I & N'_a & N'_4 \\ 0 & N_a & N_4 \end{bmatrix} \tag{7.1}$$

in which first row is for fixed head nodes. Similarly, loop incidence matrix can be written as

$$L = \begin{bmatrix} L_{41} & L_{4a} & I \end{bmatrix} \tag{7.2}$$

From the orthogonal relationship between node incidence matrix and loop incidence matrix, flow in tree elements can be written in terms of flows in chords.

$$q_t = -N_t^{-1} N_c q_c = L_t^T q_c \tag{7.3}$$

In terms of tree elements eq. 7.3 yields

$$q_1 = L_{41}^T q_4 \tag{7.4}$$
$$q_a = L_{4a}^T q_4 \tag{7.5}$$

Loop head loss equations satisfying Kirchhoff's second law are

$$h_4 + L_{41} h_1^* + L_{4a} h_a = 0 \tag{7.6}$$

where * indicates known fixed value of head loss. Equation of motion for pipes i.e. element type a are

$$h_a = m_a \frac{dq_a}{dt} + r_a q_a |q_a| = h_D + h_F \tag{7.7}$$

Subscripts D and F indicate dynamic and frictional terms. Equation for flow through valves in terms of dimensionless valve openings is given by eq. 6.10 and is restated as

$$\frac{q_4}{q_{4_o}} = \tau \sqrt{\frac{h_4}{h_{4_o}}} \tag{7.8}$$

where τ is dimensionless valve opening and subscript o indicates initial steady-state values. Equations 7.4-7.8 describe the dynamic behaviour of a system and are called state equations. The only independent variable is q_4 i.e. flow through the chords and all other variables are dependent on q_4. Hence, q_4 is the state variable.

7.3 An Optimization Model for Optimal Control of Pressure Surges

In this section, a methodology is developed for the design of optimal valve operations by determining the chord discharge i.e. q_4 using an optimisation approach. A quadratic objective function is developed for the minimisation of both head change and rate of head change at nodes. Optimisation is carried out using principles of calculus of variations.

Formulation has been carried out in dimensionless form. All variables are first transformed into dimensionless form using

$$h' = \frac{h}{h_R}; q' = \frac{q}{q_R}; t' = \frac{t}{T}; m' = \frac{m}{m_R}; \ r' = \frac{r}{r_R}; H_D = \frac{h_D}{h_R}; H_F = \frac{h_F}{h_R} \tag{7.9}$$

where subscript R stands for reference value of a variable which is different from zero. In dimensionless form eq. 7.7 becomes

$$h'_a = C_M m'_a \frac{dq'_a}{dt'} + C_F r'_a q'_a \lvert q'_a \rvert = H_D + H_F \tag{7.10}$$

where $C_M = m_R.q_R/T.h_R$ and $C_F = r_R.q_R.\lvert q_R \rvert/h_R$. In dimensionless form, nodal head matrix, H_N, can be written in terms of head losses in branches i.e. $h_a{}'$

$$N_a^T H_N = h'_a \tag{7.11}$$

7.3.1 Objective function

As frictional losses can not be controlled for given network characteristics, an objective function is developed which is based on controlling the dynamic component of nodal head and time rate of change of this head. Hence, objective of the control problem can be stated as to control the differences and rate of change of differences between actual hydraulic grade line elevations at nodes and grade line, which would exist at same discharge under steady-state. These differences and rate of change of differences are as follows:

$$H_{ND} = H_N - H_N^S = \left[N_a^T \right]^{-1} H_D \tag{7.12}$$

$$\frac{dH_{ND}}{dt'} = \frac{dH_N}{dt'} - \frac{dH_N^S}{dt'} = \left[N_a^T \right]^{-1} \frac{dH_D}{dt'} \tag{7.13}$$

In quadratic form, the objective function can be written as

$$\Phi = \frac{1}{2} \int_0^1 \left\{ \alpha . \frac{dH_{ND}}{dt'} \right\}^T \left\{ \alpha . \frac{dH_{ND}}{dt'} \right\} dt' + \frac{1}{2} \int_0^1 \left\{ \beta . H_{ND} \right\}^T \left\{ \beta . H_{ND} \right\} dt' \tag{7.14}$$

where α and β are weights assigned to each node for minimising time rate of change of head and the head itself at a particular node. Inserting value of $h_D{}'$ from eq. 7.10 and using eq. 7.5 in dimensionless form, the objective function becomes

$$\Phi = \frac{1}{2}\int_0^1\left\{\left(\frac{d^2q_4'}{dt'^2}\right)^T M_\alpha\left(\frac{d^2q_4'}{dt'^2}\right)\right\}dt' + \frac{1}{2}\int_0^1\left\{\left(\frac{dq_4'}{dt'}\right)^T M_\alpha\left(\frac{dq_4'}{dt'}\right)\right\}dt' \qquad (7.15)$$

where

$$M_\alpha = L_{4a}m_a'N_a^{-1}\alpha^2 N_a^{T^{-1}}m_a'L_{4a}^T \qquad (7.16)$$

$$M_\beta = L_{4a}m_a'N_a^{-1}\beta^2 N_a^{T^{-1}}m_a'L_{4a}^T \qquad (7.17)$$

7.3.2 Optimization model

The optimal control problem considered can be stated as: Find an optimal discharge trajectory, q_4, which minimises the objective function given by eq. 7.15 and satisfies the following boundary conditions:
At $t = 0$ i.e. $t' = 0$

$$q_4'(0) = q_{4_0}'; \left.\frac{dq_4'}{dt'}\right|_{t'=0} = 0 \qquad (7.18)$$

and at $t = T$ i.e. $t' = 1$

$$q_4'(1) = q_{4T}'; \left.\frac{dq_4'}{dt'}\right|_{t'=1} = 0 \qquad (7.19)$$

After transforming the variable dq_4/dt as

$$v = \frac{dq_4'}{dt'} \qquad (7.20)$$

and, thus, with new boundary conditions

$$v(0) = v(1) = 0 \qquad (7.21)$$

$$\int_0^1 v\,d\tau = \Delta q_4' = q_{4T}' - q_{4_0}' \qquad (7.22)$$

the Euler-Lagrange equation corresponding to eq. 7.15 is

$$M_\alpha \frac{d^2 v}{dt'^2} - M_\beta v + C = 0 \qquad (7.23)$$

where C is a column matrix of constants. Solution of eq. 7.23 is given by

$$\frac{dq_4'}{dt'} = v = X.A.\sinh \gamma t' + X.B.\cosh \gamma t' + C \qquad (7.24)$$

where γ are the eigenvalues of the matrix $M_\alpha^{-1}.M_\beta$ and X is the eigenvector matrix in which each column represents eigenvectors corresponding to each eigenvalue. A and B are column matrices of constants. Integration of eq. 7.24 gives flow in chords, q_4'

$$q_4' = X.A.\frac{\cosh \gamma t'}{\gamma} + X.B.\frac{\sinh \gamma t'}{\gamma} + Ct' + D \qquad (7.25)$$

where D is a column matrix of constants. Constants A,B,C and D are determined by using boundary conditions stated by eq. 7.18 and eq. 7.19 for q_4' and eq. 7.22. These constants are given by

$$A = X_1^{-1}\left[X_2.X^{-1}-1\right]C \qquad (7.26)$$

$$B = -X^{-1}.C \qquad (7.27)$$

$$C = \left\{X_3.X_1^{-1}\left[X_2.X^{-1}-1\right]-X_4.X^{-1}+1\right\}^{-1}.\Delta q_4' \qquad (7.28)$$

$$D = \frac{1}{2}\left\{q_{4_0}' + q_{4_T}' - (X_5+X_6).A - X_4.B - C\right\} \qquad (7.29)$$

where

$$X_1 = \left[X_{ij}\right]\left\{\sinh \gamma_j\right\}$$
$$X_2 = \left[X_{ij}\right]\left\{\cosh \gamma_j\right\}$$

$$X_3 = \left[X_{ij}\right]\left\{(\cosh \gamma_j - 1)/\gamma_j\right\}$$
$$X_4 = \left[X_{ij}\right]\left\{\sinh \gamma_j / \gamma_j\right\}$$

$$X_5 = \left[X_{ij}\right]\left\{1/\gamma_j\right\}$$
$$X_6 = \left[X_{ij}\right]\left\{\cosh \gamma_j / \gamma_j\right\} \qquad (7.30)$$

Flow rates obtained by eq. 7.25 define flow rates through the valves and resulting valve operations are determined using eq. 7.8 after evaluating h_4 that is the head loss through valves. These valve operations transfer the system from an initial steady state to a final steady state as required.

7.4   A Single Reservoir-pipe-valve System

Consider the system of a single pipe as shown in Fig. 6.2. Pipe is 1000 m long with diameter 1.0m. If reservoir head of 50m and Darcy Weisbach's friction factor, f=0.016 is taken, resulting discharge variation from an initial value of 5m³/s to zero is derived using eq. 7.25. For the case of $\alpha = 1$ and $\beta = 1$, discharge variations at valve are shown in Fig. 7.1 by curve 1. Figs. 7.2 and 7.3 show resulting valve operation and head variation at valve end, These variations for α=1 and β=4 are shown by curve 2 in these figures. Resulting valve operation rule is smooth and flows are controlled smoothly. Obviously, as β increases, q tends to follow a straight line variation and h increases at beginning and end of valve operation.

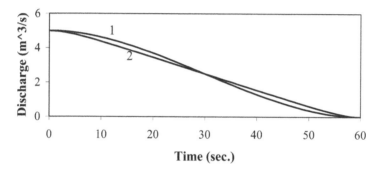

Fig. 7.1 Flow through valve

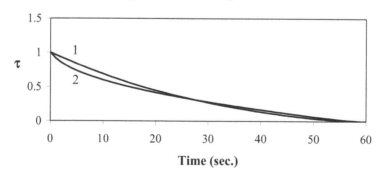

Fig. 7.2 Dimensionless valve opening

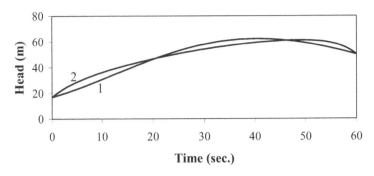

Fig. 7.3 Head at valve

In large network systems, the determination of operation rules is less straightforward. Nodes, which are located close to reservoirs generally present limited pressure variations as compared to nodes, which are distant. Also system operation may require specific limitations on maximum and minimum pressures developed at nodes as well as bounds on the time rate of change of such pressures to control the extent of water hammer occurrences. A rationale for the selection of weight factors α and β is given in next section.

7.5   Determination of Weights at Nodes

Equation 7.12 for head at node can be written as

$$H_N = \frac{m_R q_R}{T.h_R} N_a^{T^{-1}} m_a' . L_{4a}^T \frac{dq_4'}{dt'} \tag{7.31}$$

Total variation of head during the operation time becomes

$$\int_0^1 H_N(\tau)dt' = \left(\frac{m_R q_R}{T.h_R}\right).N_a^{T^{-1}} m_a' . L_{4a}^T \Delta q_4' = \varsigma \tag{7.32}$$

An objective function, which minimises time rate of change of head and head itself, can be written as

$$\Phi = \frac{1}{2}\int_0^1 \left\{ \left(\frac{dH_N}{dt'}\right)^T \alpha^2 \left(\frac{dH_N}{dt'}\right) \right\} dt' + \int_0^1 (H_N^T \beta^2 H_N)dt' + \lambda \int_0^1 H_N dt' \tag{7.33}$$

subject to boundary conditions

$$H_N(0)=0 \; ; \; H_N(1)=0 \tag{7.34}$$

where λ is the Lagrange multiplier. The Euler-Lagrange equation corresponding to eq. 7.33 is

$$\alpha^2 \frac{d^2 H_N}{dt'^2} - \beta^2 H_N - \lambda = 0 \qquad (7.35)$$

Solution of eq. 7.35 is given by

$$H_N = \frac{\varsigma \cdot \gamma}{2(\cosh\gamma - 1) - \gamma\sinh\gamma} \left\{ \begin{matrix} \sinh\gamma t' + \sinh\gamma(1 - t') \\ -\sinh\gamma \end{matrix} \right\} \qquad (7.36)$$

$\gamma = \beta^2 / \alpha^2$. It is to point out here that γ is important rather than separate values of α and β. Differentiating eq. 7.36 yields

$$\frac{dH_N}{dt'} = \frac{\varsigma \cdot \gamma^2}{2(\cosh\gamma - 1) - \gamma\sinh\gamma} \{\cosh\gamma t' - \cosh\gamma(1 - t')\} \qquad (7.37)$$

Maximum values of H_N and (dH_N / dt), which occur at $.t' = 0.5$ and $.t' = 0$, 1 respectively, are given by

$$\left. \frac{dH_N}{dt'} \right|_{max} = \pm 2\varsigma\varsigma \frac{u^2 \sinh u}{u\cosh u - \sinh u} \qquad (7.38)$$

$$H_{N_{max}} = \pm\varsigma \frac{u(\cosh u - 1)}{u\cosh u - \sinh u} \qquad (7.39)$$

where $u = \gamma / 2$. If there is a restriction of maximum time rate of change of head at a node, $(dH_N / dt)^*$, then the following inequality must hold during the time of operation of valves;

$$\left. \frac{dH_N}{dt'} \right|_{max} \leq \left. \frac{dH_N^*}{dt'} \right|_{max} \qquad (7.40)$$

which implies that

$$\frac{u^2 \sinh u}{u\cosh u - \sinh u} \leq \left. \frac{dH_N^*}{dt'} \right|_{max} \frac{1}{2|\varsigma|} = F_1 \qquad (7.41)$$

Similarly, if there is a restriction on maximum head rise/drop, then the following inequality must hold;

$$H_{N_{max}} \leq H_{N_{max}}^* \qquad (7.42)$$

which implies that

$$\frac{u(\cosh u - 1)}{u \cosh u - \sinh u} \le \frac{H_N^*}{\varsigma} = F_2 \tag{7.43}$$

Selection of γ at a node which satisfies relations given by equations 7.41 and 7.43, would keep the maximum head and maximum time rate of change of head within bounds. A plot of functions, F_1 and F_2, with respect to γ is shown in Fig. 7.4.

Function F_1 has a value of 3.0 at very low value of γ and it increases almost linearly with γ. Function F_2 has a value of 1.5 at very low value of γ and it decreases with the increase of γ and reaches asymptotically to a value of 1.0 at very high value of γ. Following inferences can be drawn for the selection of γ:

1. For a specified value of $(dH_N / dt)_{max}$,calculate F_1

    (a) If $F_1 > 3.0$, find the value of γ from the graph which corresponds to F_1. Let this value of γ be called γ^*. Any value of γ which is less than γ^* would give time rate of change of head less than the specified value.

    (b) If $F_1 < 3.0$, increase the total time of operation. The case where $F_1 < 3.0$ implies that for a given T, it is not possible to reduce dH_N / dt to the specified value.

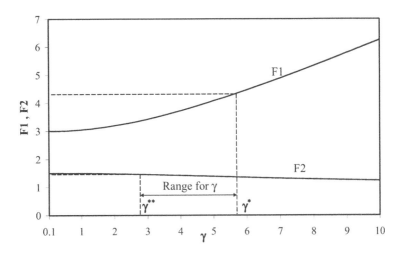

Fig. 7.4 Function F_1 and F_2 versus γ

2. For a specified value of maximum h_N, calculate F_2

    (a) If $F_2 > 1.5$, any value of γ would satisfy the specification of max. head.

(b) If $1.0 < F_2 < 1.5$, find γ which corresponds to F_2. Let this value of γ be called γ^{**}. Any value of γ which is more than γ^{**} would give head at a node less than the specified maximum head.

(c) If $F_2 < 1.0$, increase the time of operation. This case implies that for a given T it is not possible to limit the maximum head to the specified value.

Hence selected value of γ at a node should satisfy following relation

$$\gamma^{**} < \gamma < \gamma^{*}$$

If $\gamma^{**} > \gamma^{*}$, increase the time of operation, T. This case implies that for a given time of operation, T, it is not possible to satisfy the two conditions, max. H_N and max. dH_N / dt, together.

A general algorithm for the determination of valve operation rules for optimal control of pressure surges by using present methodology is given in Fig. 7.4.

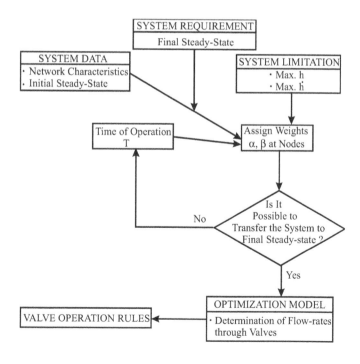

Fig. 7.5 Algorithm for the problem of optimal control of pressure surges by valve operations

7.6   Application Examples

The network system shown in Fig. 7.6 is taken as a first example. Network element characteristics are given in Table 7.1. Three cases have been considered. Table 7.2 shows the initial steady-state and final steady-states for these three cases. Value of α for each node is taken as 1.0. Value of β for nodes A, B and C is taken as 1.5 and for other nodes $\beta = 1.0$ is

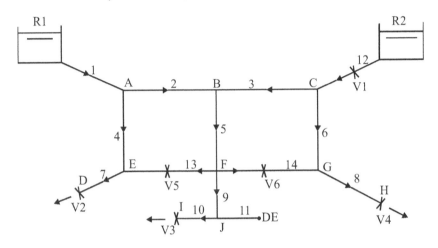

R : Reservoir
V : Valve
taken.   DE : Dead End

Fig. 7.6 A pipe network as an example I

Table 7.1 Characteristics of network shown in Fig. 7.6

Pipe	Length (m)	Diameter (m)
1	1000.	1.5
2	1500.	1.0
3	1500.	1.0
4	1000.	0.6
5	1000.	1.5
6	1000.	0.7
7	600.	1.5
8	400.	1.5
9	300.	1.0
10	400.	1.0
11	500.	1.5
12	1000.	1.5
13	1000.	1.0
14	1000.	1.0
f = .016 for each pipe		
Reservoir elevations: R1=80. m; R2=100.m		

Table 7.2 Initial and final steady-states

Pipe discharge (m³/s)	Initial steady-state	Final steady-state		
		Case I	Case II	Case III
1	4.0	0.	4.0	6.0
2	2.5	0.	2.0	4.0
3	2.5	0.	3.0	-1.0
4	1.5	0.	2.0	2.0
5	5.0	0.	5.0	3.0
6	2.5	0.	2.0	1.0
7	3.0	0.	5.0	3.0
8	4.0	0.	2.0	3.0
9	2.0	0.	2.0	0.
10	]2.0	0.	2.0	0.
11	0.	0.	0.	0.
Valve Discharges (m³/s)				
V1	5.0	0.	5.0	0.
V2	3.0	0.	5.0	3.0
V3	2.0	0.	2.0	0.
V4	4.0	0.	2.0	3.0
V5	1.5	0.	3.0	1.0
V6	1.5	0.	0.0	2.0
$\tau 1$	1.	0.	1.475	0.00
$\tau 2$	1.	0.	4.035	2.85
$\tau 3$	1.	0.	0.963	0.00
$\tau 4$	1.	0.	0.348	0.69
$\tau 5$	1.	0.	1.300	0.50
$\tau 6$	1.	0.	0.000	4.14
Head at nodes (m)				
A	77.21	80.	77.21	73.71
B	64.81	80.	69.29	41.90
C	77.21	80.	87.17	39.80
D	37.98	80.	6.48	4.66
E	38.92	80.	9.11	5.60
F	60.45	80.	64.93	40.35
G	28.00	80.	55.70	31.99
H	26.88	80.	55.43	31.37
I	56.74	80.	61.23	40.38

Case I. In this case flow through all outflow nodes is reduced to zero in 60 seconds. Valve operations obtained from the developed methodology are shown in Fig. 7.7. Flow and nodal head variations are shown in Figs. 7.8-7.10. System is smoothly transferred to final steady-state without any residual transients.

Case II. In this case, it is desired to increase the outflow through valve 2 to 5.0 m³/s and to decrease the outflow through valve 4 to 2.0 m³/s while maintaining a constant outflow through valve 3. System is required to reach final steady-state in 60 seconds. Valve operations obtained are given in Fig. 7.11. Figs. 7.12-7.14 show flow and nodal head variations for this case.

Case III. In this case, flow through valve 4 is to be reduced to 3.0 m³/s. Flow through valve 2 is to be kept constant and valve 3 is to be closed completely. Moreover, it is desired to supply water only from reservoir 1 and, thus, closing valve 1 completely. Final steady-state is to be reached in 60 seconds. Using the developed methodology, system is transferred to final steady-state without any residual transients. Figures 7.15-7.18 show valve operations, flow and nodal head variations.

In all cases, full controllability of the network is guaranteed and optimal valve operations required to transfer the system to a final steady-state were obtained using the present methodology.

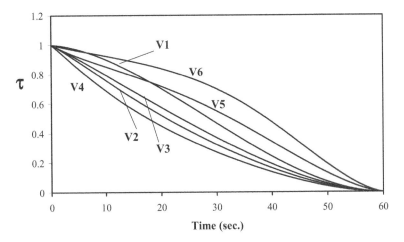

Fig. 7.7 Dimensionless valve openings (Case I)

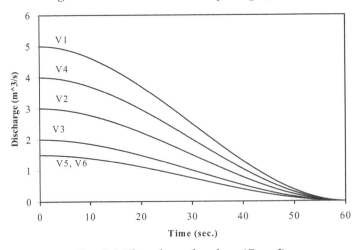

Fig. 7.8 Flow through valves (Case I)

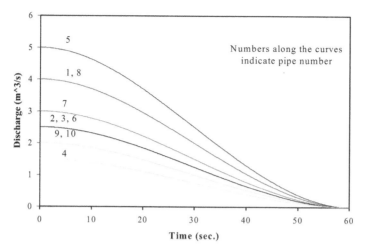

Fig. 7.9 Flow through pipes (Case I)

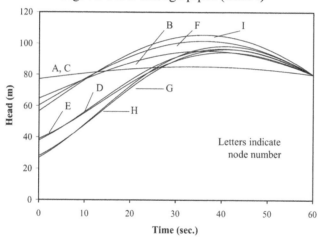

Fig. 7.10 Head at nodes (Case I)

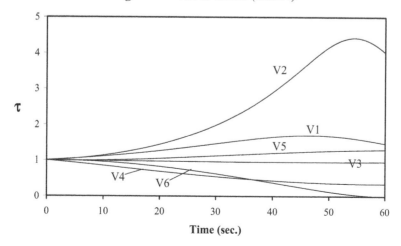

Fig. 7.11 Dimensionless valve openings (Case II)

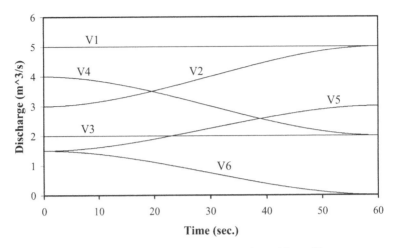

Fig. 7.12 Discharge through valves (Case II)

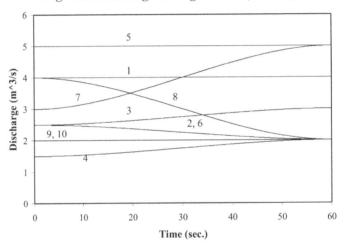

Fig. 7.13 Flow through pipes (Case II)

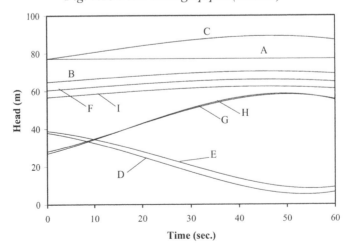

Fig. 7.14 Head at nodes (Case II)

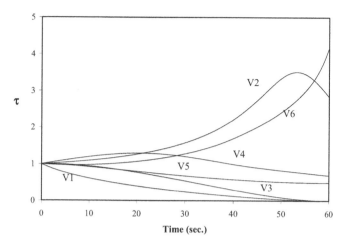

Fig. 7.15 Dimensionless valve openings (Case III)

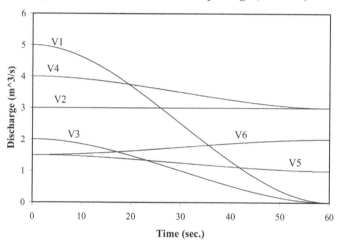

Fig. 7.16 Flow through valves (Case III)

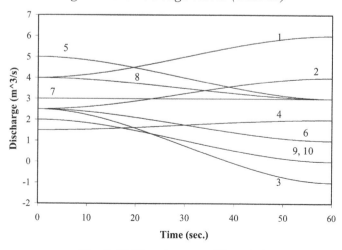

Fig. 7.17 Flow in pipes (Case III)

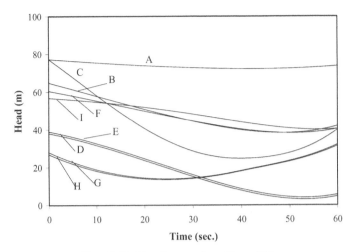

Fig. 7.18 Head at nodes (Case III)

Pipeline system shown in Fig. 7.19 is taken as example II. Shimada (1989) took this example to obtain valve operations using optimal control theory. The characteristics of the system are given in Table 7.3. Reservoir head at node 5 is 30 m and at node 6 reservoir head is 18 m. Overflow surge tank 1 at node 1 has a area of 10 m^2 with crest level at 24 m. Area and crest level of overflow surge tank 2 at node are 10 m^2 and 18 m respectively.

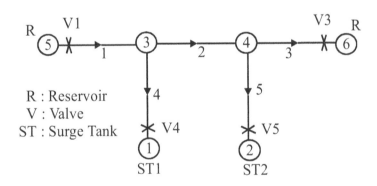

Fig. 7.19 Pipeline system for example II

Table 7.3 Characteristics of pipeline system shown in Fig. 7.19

Pipe	1	2	3	4	5
Length (m)	4000	3500	2200	1500	1300
Diameter (m)	1.5	1.2	1.0	1.0	0.8
f	.0203	.0252	.0272	.0284	.0282

Initial steady-state flow in pipes 3, 4 and 5 is 0.6 m³/s and is to be reduced to a final steady-state value of 0.4 m³/s each. Total time of operation considered is 60 seconds. Valve operations required to accomplish the job are given in Fig. 7.20. Figure 7.21 shows flow variations in pipes.

In fact this system has one extra valve, V1. Three valves, V3, V4 and V5 are sufficient to guarantee full controllability of the network. Valve V1 is taken as a branch of the spanning tree if node 5 is considered as reference node and an extended network is constructed.

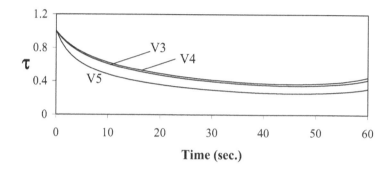

Fig. 7.20 Dimensionless valve openings

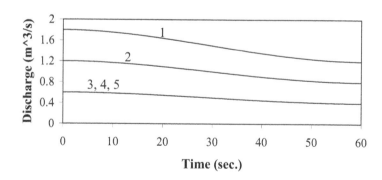

Fig. 7.21 Flow through pipes

7.7   Concluding Remarks

In pipe networks, problem of control of pressure surge by valve operations requires transfer of the system from an initial steady-state to a desired final steady-state within specified time without any residual transients. However, for full controllability of a network, number of valves should be equal to the number of chords in the system. Moreover, locations of these valves are in a co-tree of the system. Otherwise, system can not be fully controlled and residual transients will exist at the end of valve operation.

Optimal control of pressure surges requires minimum change in pressure head in the system and a smooth transfer of the system from an initial steady-state to a final steady-state. Valve operations guaranteeing these optimal criteria are called optimal valve operations. In this chapter, a methodology for the design of optimal valve operations by using principles of calculus of variations has been presented. Proposed procedure can well be termed as valve stroking for pressure surges. In the developed mathematical optimisation model, the boundary conditions guarantee satisfaction of two Kirchoff' Laws with q, h & operational controls reaching their final steady state values at the end of valve operational time. The first derivatives of q and h also become zero at the end valve operations. Though, second and higher order derivatives of q and h are not zero at the end of valve operational time, but the same can be neglected. This is precise for practical engineering applications.

8 | Transient System Component Design for Pressure Surges

Conventional methods for evaluation of design parameters of system components specially used for controlling transients such as surge tank, air vessel, valve operations, conveying element etc. are based on a sequence of trials and errors. Typical procedures consist of assuming a component design and analysing the result of transients due to different planned and accidental cases. If the resulting behaviour of the system is not satisfactory, a new design of the component is adopted. Procedure is repeated until established criteria are satisfactorily met. These procedures are tedious especially for large and complex networks where much iteration is required to determine the design parameters of a component.

In this chapter, an approach to transient system component design is developed which determines design parameters of the components directly thus, avoiding sequence of trials and errors. Application of the developed approach is shown for the design of surge tank and valve operations. This approach shows many applications in the field of system design for the specified transient behaviour.

8.1   A General Algorithm for Transient System Component Design

To control pressure surges in networks, transient-controlling elements like surge tank, air vessel, conveying element etc are generally used. These system components are designed to keep pressure surges under specified limits. Available methodologies for the design of such components are based on the principles of 'analysis' rather than 'synthesis'. These procedures require number of trial and errors to finalise the design parameters of such components, which produces the desired transient behaviour in the system. Development of methodologies, which determine the design parameters of some network components directly for the specified transients in the system are essential for avoiding tedious iterative procedures.

The present methodology is based on network synthesis approach. The principle behind the present methodology is that the transient behaviour in some parts of the system is specified and some parts of the system are kept undetermined to be designed in order to meet the imposed specifications. Analysis of the system directly yields the transient behaviour of these parts, which are then accordingly designed. The procedure can be well described as transient synthesis.

In the network, components to be designed are modelled as elements type 4. These components can also be modelled as elements type 1 or 2, if head loss or flow through these components is specified. Components with specified transients are modelled as elements type 3. A network model is constructed which is determined in nature i.e. number of elements type 4 are equal to the number of elements type 3. Distribution of different types of elements in the network should obey necessary and sufficient conditions for the existence and uniqueness of a solution as described in chapter 4.

Network model is represented by a set of equations as described in section 6.4 of chapter 6. Analysis of the system is carried out for the specified transients i.e. boundary conditions. This provides flow and head loss behaviour in elements type 4 explicitly. These results can be optimised to get design parameters of the desired system component out of element type 4. Formulation of optimization model depends on the type of the component to be designed and its desired characteristics.

A second analysis of the system is carried out by considering now elements type 4 and 3 as type 5 elements which provides the transient behaviour in the elements considered as type 3 elements in first analysis. Difference between the transient specifications in elements type 3 and the transient response or modified specifications obtained by this second analysis depends on the efficacy of the optimization model.

The algorithm for transient system component design is shown in Fig. 8.1 and can be summarised in the following steps:

(i)   Model the system components in nodes/elements type 1,2,3,4 and 5.
(ii)   Construct a closed network.
(iii)   Check for the problem solvability i.e. existence and uniqueness of a solution.
(iv)   Put initial conditions and boundary conditions.
(v)   Analyse the system by numerical integrating a set of ordinary differential equations consisting of independent variables q_b , the flow in type b elements and z, water level in surge tanks.
(vi)   Obtain the design parameters of a system component by optimizing the head loss and flow in element type 4 to suit a desired component behaviour.

The algorithm for transient system component design can be used for a variety of applications in the field of water distribution networks, pumping stations, pipelines, hydropower systems etc. Design of surge tank in a hydropower system for the specified transients at turbine, design of air vessel in a pumping station, design of valve operation measures for the specified transients etc. are some examples of such application.

In the following sections, application of the algorithm has been shown for the design of surge tank and valve operation measures.

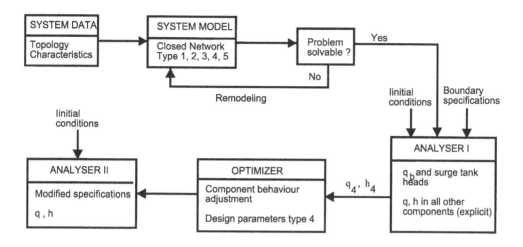

Fig. 8.1 Algorithm for transient system component design

8.2   Application to Surge Tank Design

Surge tanks in a network system are used to control transients. Design parameters of a surge tank are evaluated to keep the transients arising due to various planned and accidental operation cases within limits. Conventional methods for their design use iterative procedure based on trial and errors. Using the present algorithm, design parameters of a surge tank can be determined directly.

A surge tank with unknown parameters is considered as type 4 element. The desired transient response of some other element is specified and the element is modelled as type 3 element. From the analysis of the network, discharge, q_4 , and head loss, h_4 , are calculated explicitly. These results are optimized to obtain design parameters of the surge tank.

8.2.1   Optimization model

If z is the water level in the surge tank, then

$$h_4 = r \, q_4 \cdot |q_4| + z \qquad\qquad (8.1)$$

r is the head loss coefficient of a throttled surge tank. As the q_4 and h_4 obtained from the analysis may not show a true behaviour of a surge tank, a assumed discharge, q, is added which may make a type 4 element as a surge tank. Thus,

$$q + q_4 = A_s \frac{dz}{dt} \q\qquad\qquad (8.2)$$

From equations 8.1 and 8.2,

$$q = A_s \frac{dh_4}{dt} - 2 r A_s |q_4| \frac{dq_4}{dt} - q_4 \qquad (8.3)$$

Equation 8.3 contains two variables A_s and rA_s. In order to achieve that type 4 element behaves as closely as possible as a surge tank, a minimization problem is formulated with the objective function being:

$$\min \phi = \int_0^T q^2 \, dt \qquad (8.4)$$

Minimization of function ϕ with respect to A_s and rA_s yields two simultaneous equations

$$c_1 A_s - 2 c_2 (rA_s) - c_3 = 0 \qquad (8.5)$$

$$c_2 A_s - 2 c_4 (rA_s) - c_5 = 0 \qquad (8.6)$$

where

$$c_1 = \int_0^T \left(\frac{dh_4}{dt} \right)^2 dt \qquad (8.7)$$

$$c_2 = \int_0^T |q_4| \frac{dq_4}{dt} \frac{dh_4}{dt} \, dt \qquad (8.8)$$

$$c_3 = \int_0^T q_4 \frac{dh_4}{dt} \, dt \qquad (8.9)$$

$$c_4 = \int_0^T |q_4|^2 \left(\frac{dq_4}{dt} \right)^2 dt \qquad (8.10)$$

$$c_5 = \int_0^T q_4 |q_4| \frac{dq_4}{dt} \, dt \qquad (8.11)$$

From equations 8.5 and 8.6, area and throttling coefficient of a surge tank may be calculated. In the above formulation, only two parameters i.e. area and throttling coefficient have been taken. Other parameters such as expansion chamber size may be included in the formulation.

8.2.2   Examples of application in the field of hydropower systems

In hydropower system, design parameters of a surge tank are determined to keep oscillations within limits and, thus, keeping the head rise/drop at turbine within specified limits. As per current methodology, parameters of a surge tank are designed by trial and error procedure. Advantages of the present algorithm are shown by following examples in which design parameters of a surge tank are evaluated directly and, thus, avoiding trial and error procedure. In the following examples, surge tank with unknown dimensions has been considered as type 4 element and turbine as type 3 element with specified transients.

(i)   Consider a single surge tank hydropower system as shown in Fig. 8.2. Let the gross head and maximum discharge through turbine be 100 m and 100 m³/s. For the linearised frictionless case, it can be shown that the head at turbine for a sudden opening of wicket gates is given by

$$h_T = a.\sin \omega t \qquad (8.12)$$

where a = q_T. $\sqrt{}$ (m/A_s) and ω^2 = 1/(m. A_s) with m = 1 / (A.g). A and l are the cross sectional area and length of the head race tunnel and a is the maximum amplitude of oscillations. If a maximum allowable variation of net turbine head is 22.6 m, A_s = 100 m² and ω = 0.0443 rad/s. Applying a head variation at turbine given by eq. 8.12 with the calculated values of a and ω and a linear discharge variation from zero to 100 m³/s in 30 seconds and then keeping it constant, analysis of the system is carried out which provides q_4 and h_4 explicitly. The calculated values of the coefficients given by eqs. 8.7 to 8.11 are c_1 = 407 ; c_2 = -25885 ; c_3 = 35701 ; c_4 = 88348203 and c_5 = 51190. Solving eqs. 8.5 and 8.6 provides area of surge tank equal to 89.4 m² with throttling coefficient equal to zero. In the analysis, a Darcy-Weissbach's friction factor f = 0.016 is taken for the tunnels.

Analysis of the system with surge tank area of 89.4 m² shows a maximum drop of head at turbine equal to 22.5 m.

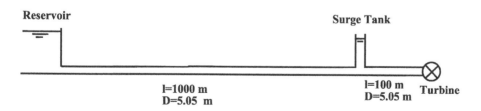

Reservoir   **Surge Tank**

l=1000 m
D=5.05 m

l=100 m
D=5.05 m   **Turbine**

Fig. 8.2 A hydropower system for example (i)

(ii)   Consider the hydropower system with two surge tanks as shown in Fig. 8.3. Let the area of the downstream surge tank be given or can have a certain maximum value due to some reasons. Upstream surge tank is to be designed and is thus considered as element type 4. Turbine is modelled as element type 3 with specified time histories of discharge and head loss. The head at turbine for a linearised frictionless case for a sudden opening of wicket gates will be

$$h_T = -a_1 \sin \omega_1 t - a_2 \sin \omega_2 t \qquad (8.13)$$

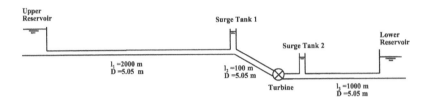

Fig. 8.3 A hydropower system for example (ii)

where $a_1 = q_T \cdot \sqrt{(m_1/A_{s1})}$; $a_2 = q_T \cdot \sqrt{(m_3/A_{s2})}$; $\omega_1^2 = 1/(m_1 A_{s1})$ and $\omega_2^2 = 1/(m_3 A_{s2})$ with $m_1 = l_1/(A_1.g)$ and $m_3 = l_3/(A_3.g)$. Gross head at turbine is 100 m and the maximum discharge through wicket gates is 100 m³/s. For assumed $A_{s2} = 100$ m², $a_2 = 22.6$ m and $\omega_2 = 0.0443$ rad/s. if the maximum head variation of 50 m is allowed at turbine, $a_1 \approx 27.4$ m which gives $A_{s1} = 136$ m² and $\omega_1 = 0.0268$ rad/s. A head variation given by equation 8.13 with the calculated values of a_1, a_2, ω_1 and ω_2 is applied at turbine for 300 seconds. A linear increase in discharge from zero to full value in 30 seconds is considered through the turbine. With the above specifications at turbine, analysis of the system provides the following values of the coefficients for the type 4 surge tank: $c_1 = 131.6$; $c_2 = -20115$; $c_3 = 11556$; $c_4 = 14292136$; $c_5 = 611886$. Using equations 8.5 and 8.6, area of the surge tank is found to be equal to 120 m² with r=0. Simple analysis of the system with $A_{s1} = 120$ m² and $A_{s2} = 100$ m² shows a maximum head drop of 48.3 m at turbine.

In the above examples, only the case of load acceptance is considered. However, other cases such as load rejection, load acceptance followed by load rejection etc. should be considered. In the examples, surge tank with unknown dimensions is considered as type 4 element. However, if a surge tank with limited amplitude of oscillations has to be designed, it can be considered as type 1 element with specified head loss and turbine can be modelled as type 5 element with known power output. In this case, discharge through the wicket gates will vary to meet the specified power, for the case when a perfect regulator works. Design parameters of the surge tank can be evaluated after optimising the results obtained from the analysis giving explicitly the flow through element type 1 i.e. surge tank.

8.3   Application to Design of Valve Operations

Problem of design of valve operations for the specified transients has been considered in the chapters 6 and 7. Consider a determined problem in which number of element type 3 are equal to the number of element type 4. Elements type 4 are valves and type 3 elements are those in which transients are prescribed. As shown in section 6.4, for the case of partial control problems i.e. for the networks in which not every loop is having a controlling valve, valve operations obtained from the analysis extend over a the entire duration of transients. Though the transient specifications are met by these valve operations but the long operation of these valves is generally not desirable. Residual transients in these partial control problems can not be ruled out, however, design of valve operations for the specified time can be achieved by optimising a selected objective function.

8.3.1   Optimization model

As stated earlier, a valve with unknown coefficient is considered as a type 4 element. Network analysis yields discharge, q_4 , and head loss, h_4 and thus, valve coefficients are determined directly using equation 6.10 restated as below

$$q_v = \tau q_{v0} \sqrt{\frac{h_v}{h_{v0}}} \qquad (8.14)$$

Let these coefficients be called τ_4^c. These valve operations exist until residual transients persist. If the valve operations are desired to be completed at the time $t = T$, the actual valve openings may be represented by a Fourier series as follows:

$$\tau_4 = \tau_0(1 - t/T) + \tau_f \, t/T + \sum_{n=1}^{\infty} a_n \sin(n\pi t/T) \qquad (8.15)$$

where τ_0 and τ_f are valve openings corresponding to initial and final steady state. In order to achieve that τ_4 is as close as possible to τ_4^c, a minimisation problem is formulated with the objective function being

$$\min \phi = \int_0^T (\tau_4^c - \tau_4)^2 \, dt \qquad (8.16)$$

Parameters a_n are determined using the condition $\partial\phi/\partial a_n = 0$ which gives

$$a_n = 2\frac{(I_{1n} - I_{2n} - I_{3n})}{T} \qquad (8.17)$$

where

$$I_{1n} = \sum_{N=0}^{T/\Delta t} \tau_4^c \sin\frac{n\pi N\Delta t}{T} \Delta t \qquad (8.18)$$

$$I_{2n} = \frac{\tau_0 T}{n\pi} \qquad (8.19)$$

$$I_{3n} = \frac{\tau_f T}{n\pi} \cos n\pi \qquad (8.20)$$

Knowing the values of τ_4^c, a_n is calculated and then τ_4 is determined using equation 8.15 for the specified time of valve operations T.

8.3.2 Example of application to pipe network

Network shown in Fig. 8.4 is taken as an example for which the system parameters are given in Table 8.2. Using the present methodology the system is transferred from an initial steady-state to a final steady-state as given in Table 8.1.

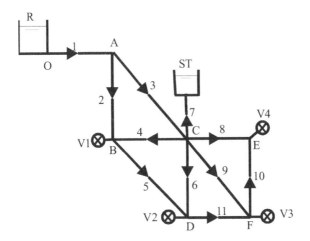

V : valve
R : reservoir
ST : surge tank

Number along the edges shows pipe number

Fig. 8.4 A pipe network

Table 8.1. Initial & Final Steady-States : Discharges, Heads and Valve Openings

Pipe	Q (m³/s)		Node	Head (m)		Valve	q (m³/s)	
	Initial	Final		Initial	Final		Initial	Final
1	25.0	8.14	O	50.0	50.0	V1	5.0	1.23
2	4.03	1.27	A	45.19	49.49	V2	10.0	3.03
3	20.97	6.87	B	17.41	46.75	V3	4.0	1.99
4	5.80	1.78	C	35.30	48.43	V4	6.0	1.89
5	4.83	1.81	D	11.31	45.89		Valve opening (τ)	
6	8.20	2.67	E	4.55	44.76			
7	0.0	0.0	F	7.32	44.97	V1	1.0	0.15
8	3.21	1.11				V2	1.0	0.15
9	3.75	1.32				V3	1.0	0.20
10	2.79	0.78				V4	1.0	0.10
11	3.03	1.45						

Valve 1 and 3 have been taken as type 3 element in which a transient behaviour in terms of linear increase of head loss from initial steady-state value to final steady-state value in 30 seconds and linear valve closure, τ_3=1 to .15 for valve 1 and τ_3=1 to .0.2 for valve 3, in 30 seconds are specified. System is analysed and resulting τ_4^c, flow and head variations in pipes are shown in Fig. 8.5 to 8.7. τ_4^c are optimised to get valve

Table 8.2. Characteristics for Pipe Network in Fig. 8.4

Pipe	1	2	3	4	5	6	7	8	9	10	11
l (km)	1.0	1.0	1.5	1.0	1.0	1.5	.05	1.0	1.5	1.0	1.0
D (m)	2.8	0.95	2.45	1.2	1.5	1.3	1.0	0.85	1.0	1.3	1.25
$f = 0.016$ for all pipes ; Area of surge tank $= 40$ m^2											

operation rules, τ_4, for the specified operation time of 30 seconds. Resulting τ_4 are shown in Fig. 8.5. System is again analysed using optimised τ_4 and specified τ_3. Figs. 8.6 and 8.7 also show the resulting flow and head variations in pipes. System is smoothly transferred to final steady-state without violating much the specified transient behaviour in element type 3 i.e. valve1 and 3.

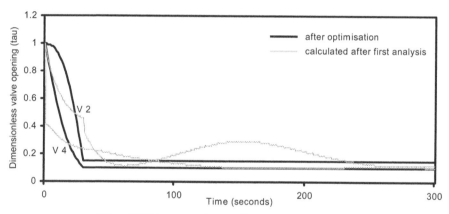

Fig. 8.5 Dimensionless valve operation rules

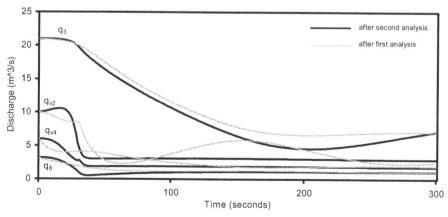

Fig. 8.6 Discharge variation in pipes and valves

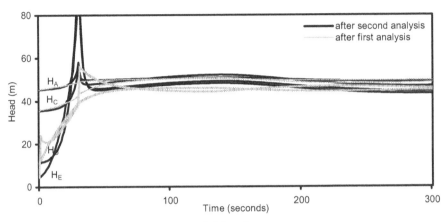

Fig. 8.7 Nodal head variations

8.4   Concluding Remarks

Design of system components such as surge tank, air vessel, valve operations, pipe etc is done to keep the transients within acceptable limits. Conventional procedures for their design are based on analysis approach rather than synthesis. Design procedures are based on tedious trial and error method. In this chapter, a general algorithm is developed which provides design parameters of the components directly and, thus, avoiding trial and error procedure.

In a hydraulic network, some components are designed in such a way that provides a desired transient behaviour in other components or at some locations. The principle behind the developed methodology in this chapter is that the transient behaviour in some parts of the system is specified and some other parts of the system are kept undetermined to be designed in order to meet the imposed specifications. Analysis of the system provides transient behaviour of these parts which are then accordingly designed.

The base of the developed algorithm is the presence of two spanning trees in the network model, T_{13a} and T_{14a}, as described in chapter 4. Element type 4 is considered as a network component to be designed. Analysis of the system for the prescribed transients yields transient behaviour of this element which can be suitably optimized to get a desired system component out of this element.

From the examples, it is evident that after second analysis i.e. after designing the desired component, the transient specifications are not met exactly. This is generally unavoidable because transient behaviour of element type 4 obtained after first analysis can not depict the transient behaviour of the desired component exactly. However, error can be minimised by specifying transients in type 3 elements which are as close to reality as possible. Further research is required in the development of optimization model which minimises this error.

The developed methodology has variety of applications in the field of design and operation of distribution systems, hydropower systems, pumping stations, industrial piping systems etc.

9 | Control of Transients Using Elastic Model

Design of valve operations to control transients is a synthesis approach in which the variations of boundary conditions are determined to obtain a desired system response. The pioneering work in the development of this approach was carried out by Streeter (1963, 1967). The procedure developed by Streeter consists of design of valve operations for the specified head variations at valves with no residual transients at the end of valve operations and is termed as valve stroking. The technique was further developed by Propson (1970) who developed procedures for valve stroking with specified time or maximum head rise/fall including friction. The technique has also been applied to gas pipelines (Stoner, 1968) and open channel surges (Wylie, 1969). However, development of this technique is still restricted to some simple piping systems.

Though the analysis of transients in complex systems using elastic model can be carried out precisely with minimal efforts by using computer packages, the problem of design of valve operations for transients control is not straightforward. However, because systems are becoming more complex, this has led to new surge of interest in improving the efficiency and accuracy of simplified model i.e. rigid model. Rigid model can be applied to a system if the elastic model produces similar transients or within acceptable error limits. This requires an investigation into the scope of rigid model versus elastic one. Some studies have been carried out in this area (Parmakian, 1955; Valentine, 1965; Wood, 1973; Karney, 1985, 1990 among others) which define the limit of applicability of rigid model in terms of Allievi's dimensionless parameters, ρ and θ. These studies are of a very restricted nature in that the majority have assumed a linear valve closure and frictionless flow. Shape of the prescribed discharge law at the valve end and frictional effects also influence the applicability of the rigid model.

This chapter concentrates first on a simple hydraulic system. The basic simplified equations and boundary behaviour are first rendered dimensionless using dimensionless parameters to predict the error between elastic and rigid models for the different prescribed discharge laws. Effects of friction have also been investigated. Guidelines are presented in the form of graphs and equations. Next, valve stroking case for single pipe is revisited and the same is discussed for more complex systems. In the end, the case of transient control in a single pipe system with the permitted residual transients is investigated using synthensis approach.

9.1   Pipe Flow Basic Equations and Dimensionless Parameters

For the inertial dynamic model that considers elastic effects, known as water hammer, the basic equations of continuity and motion governing one dimensional flow in a prismatic pipe may be written in the simplified form after neglecting slope and convective terms, respectively, as follows (Wylie and Streeter, 1993):

$$\frac{gA}{a^2}\frac{\partial H}{\partial t} + \frac{\partial q}{\partial x} = 0 \tag{9.1}$$

$$\frac{\partial q}{\partial t} + gA\frac{\partial H}{\partial x} + f\frac{q|q|}{2DA} = 0 \tag{9.2}$$

The dependent variables are the piezometric head H and discharge q; x and t are the independent variables denoting distance along pipeline and time respectively; and a is the wave speed. All other variables have the same nomenclature as previously defined. This set of equations is usually solved by the method of characteristics to provide a numerical representation of functions $q = q(t, x)$ and $H = H(t, x)$ for a set of boundary conditions.

Now, if either $a \rightarrow \infty$ (physically tending to a real rigid fluid pipe system) or $\partial H/\partial t \rightarrow 0$ (the elastic system behaving as a rigid one) the continuity equation reduces to

$$\frac{\partial q}{\partial x} = 0 \tag{9.3}$$

so that flow becomes uniform. The energy equation 9.2 can be integrated to give

$$H_u - H_d = \frac{L}{gA}\frac{dq}{dt} + \frac{fL}{2gDA^2}q|q| \tag{9.4}$$

where H_u and H_d are upstream and downstream pipe heads respectively. Equation 9.4 states that the instantaneous head difference results from both pipe friction and flow acceleration and, in fact, is the equation of motion for pipe considering rigid model.

To enable comparisons between models and a generalised parametric study, a new set of reduced variables is defined, by selecting reference values of discharge (velocity), head, length and time. These values are as follows:

$$v' = \frac{q}{q_R} = \frac{v}{v_R}; \quad H' = \frac{H}{H_R}; \quad x' = \frac{x}{L}; \quad t' = \frac{t}{T_w} \tag{9.5}$$

where q_R (v_R) = initial (maximum expected) discharge (velocity); H_R = static driving head (see Fig. 9.1); and T_w = water starting time Lv_R / gH_R. Physically, T_w is the time required to accelerate a rigid frictionless column in a pipe from 0 to q_R, or to stop the same liquid mass from q_R to zero, when driven by a constant pressure head H_R. Thus T_w is seen to be only associated only with the fluid inertia.

Rewriting Eqs. 9.1 and 9.2 in terms of the above dimensionless variables yields

$$\frac{\partial H'}{\partial t'} + (2\rho)^2 \frac{\partial v'}{\partial x'} = 0 \qquad (9.6)$$

$$\frac{\partial v'}{\partial t'} + \frac{\partial H'}{\partial x'} + h'_f v' |v'| = 0 \qquad (9.7)$$

which contain Allievi's elastic parameter ρ and friction parameter h'_f given, respectively, as

$$\rho = \frac{a v_R}{2 g H_R}; \quad h'_f = \frac{h_f}{H_R} \qquad (9.8)$$

Allievi's parameter represents the ratio of Joukowsky overpressure $a v_R/g$ to twice the system head reference H_R or, alternatively, the ratio between the water starting time $T_w = l v_R/g H_R$ and the elastic wave travel time $T_e = 2l/a$. The term h_f is the frictional head loss.

Physically, ρ can be viewed as a measure of the relative importance of inertial and elastic effects. In general, when ρ is much greater than unity, the inertia (acceleration) forces, due to the unsteadiness of the flow, dominate and it may be justifiable to neglect elastic effects and use the rigid model. However, this can only be viewed as general guidance because ρ dose not consider the boundary conditions. As the rigid model is characterized by constant discharge along the pipe element at each instant, therefore as

$$\frac{\partial v'}{\partial x'} \rightarrow 0 \qquad (9.9)$$

from eq. 9.6

$$\frac{1}{(2\rho)^2} \frac{\partial H'}{\partial t'} \rightarrow 0 \qquad (9.10)$$

which shows the two ways for the rigid model to apply: either combined pipe water elasticity makes ρ very large or boundary conditions changes slowly enough to make $\partial H'/\partial t'$ negligible.

Another important aspect is posed by the boundary conditions. Pipe pressure variations are the result of flow changes at the boundary. In general, a boundary condition is modelled by a set of equations relating the problem dependent variables, H and q, with new variables introduced by the boundary element itself and possibly time t. Hence, parameters characterizing the boundary conditions are required. The discharge/head equation at the boundary and one associated characteristic time can characterize the boundary conditions. An obvious choice for the charateristic time is the total time of forcing function, T i.e. total time of valve operation. When T is compared with wave time scale $T_e = 2l/a$ or the inertia time scale T_w (water starting time), two new dimensionless parameters are obtained

$$\theta = \frac{T}{T_c} = \frac{aT}{2l} \; ; \quad \Gamma = \frac{T}{T_w} = \frac{gAH_RT}{l\,q_R} \tag{9.11}$$

Note that θ and Γ are related as

$$\theta = \rho\Gamma \tag{9.12}$$

Hence, either of the two parameters may be used as the time scale of the boundary condition.

For the applicability of rigid model, the case of linear valve closure (i.e. uniform reduction of open area with time) has been extensively studied. Parmakian (1955) stipulated that, in the case of linear valve closure, the rigid model gives correct results for the pressure rise if

$$T > \frac{l}{1000} \tag{9.13}$$

holds, where T is the closure time in seconds and l is pipe length in feet. This equation emphasizes the decreased importance of considering the elastic effects as changes at the boundary occur more slowly relative to the conduit length.

Valentine (1965) presented graphically a quantitative representation of the disagreement between rigid and elastic models suggesting the limiting criteria: $\theta \geq 8$ and $\rho \geq 1.2$. In the discussion of Valentine's paper, O'Meill and Graze (1966) showed that these limits are extremely conservative except for high values of ρ when significant error is involved. Also Stephenson (1966) compared two models and proposed $\theta > 10$ as a criterion. In a more refined study, Wood (1973) presented a complete set of charts and found the region of validity to be given by $\rho \geq (\theta - 2.85)/(\theta - 3)$ for $\rho < 3$ and $\rho \leq 0.3\,\theta^2 - \theta + 3$ for $\rho > 3$. More recently, Karney and Ruus (1985) suggested as a rough criterion the use of $\theta = 5$.

These studies are very restricted in nature as only linear valve closure case and frictionless flow were considered. Influence of different prescribed discharge laws and effects of friction are to be investigated. As shown in the next section, the transient behaviour depends not only on the speed of variation of flow velocity but also on the shape of the discharge-time (or H-t) diagram at the boundary.

9.2   Error Between Elastic and Rigid Models for a Single Reservoir-Pipe-Valve System

The discussion in this section concentrates on estimating the magnitude of pressure changes following a valve closure. For the basic installation shown in Fig. 9.1, it is assumed that water level in the reservoir remain constant during the transient conditions. To define the applicability of the rigid model results are presented by a set of dimensionless plots where the effects of friction, inertia and the elasticity on the maximum pressure head are independently evaluated. These plots predict the rigid and elastic responses for each case and, as a result, the error in rigid column theory.

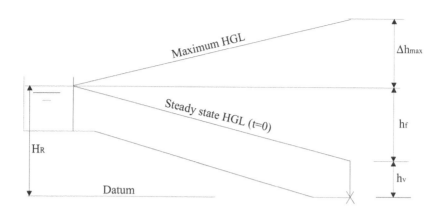

Fig. 9.1 A basic reservoir-pipe-valve system

Two types of circumstances are found that limit the applicability of the rigid model: (1) Very rapid changes generating important elastic effects, in which the rigid model gives higher values for peak pressures: and (2) cases in which fluid compressibility and pipe wall elasticity is absolutely essential regardless of the rate the discharge changes, as is the case of boundary conditions with initial discontinuous gradients in H and/or q. This latter cases may generate higher relative errors.

First, a extreme case of instantaneous valve closure is considered and then the influence of prescribed discharge laws.

9.2.1   Case of instantaneous closure of valve

Instantaneous flow stoppage of an incompressible fluid in an inelastic pipe causes an infinite surge pressure. However, because in real cases both the pipe and fluid have small but nonzero elasticity and compressibility, the surge pressure generated is finite. Therefore, rapid changes in flow rate obviously correspond to the zone where head variations given by the rigid odel are not correct and exceed the elastic model results.

For instantaneous closure case, from eqs. 9.6 and 9.7, the transient behaviour is only governed by pipe parameters ρ and h'_f. Neglecting friction effects ($h'_f = 0$), the elastic model gives the dimensionless maximum head change given by the Joukowsky equation

$$\Delta h' = 2\rho \qquad\qquad (9.14)$$

For frictional flow, Fig. 9.2 represents, for different values of h'_f values, the numerical solution of eqs. 9.6 and 9.7 when friction is included. The different curves for maximum $\Delta h'$ collapse into a single curve for $\rho \geq h'_f$; that is, friction influences the maximum pressure rise only when $\rho < h'_f$ (i.e when elastic effects play a significant part in determining transient behaviour.

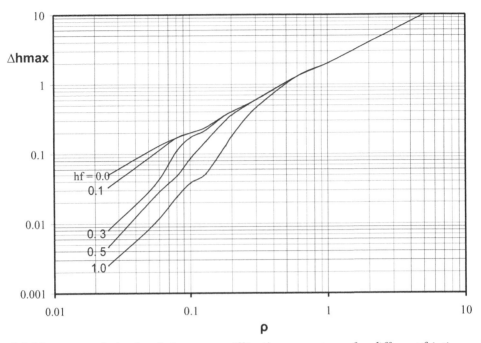

*Fig. 9.2 Maximum relative head rise versus Allievi's parameter ρ for different friction values
(sudden valve closure case)*

9.2.2   Influence of prescribed discharge laws

A boundary condition in which a law specifies q versus t is called a flow source. Flow
sources generate transients that generally depend not only on the speed of the flow variation
but also the shape of the law. Two types of discharge laws are considered.

First discharge law

The first discharge law considered is given by

$$q = q_o \left[1 - \left(\frac{t}{T} \right)^s \right] \tag{9.15}$$

where the exponent s ≥ 1 is the shape factor. Fig. 9.3 shows the variation of discharge for s =
1,2 and 3.

Rigid model theory gives the pressure change from combining eqs. 9.4 and 9.15 as

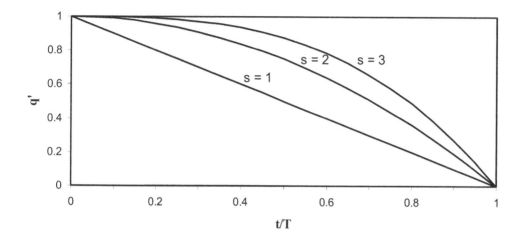

Fig. 9.3 Typical discharge profiles for s = 1, 2 and 3

$$\Delta h' = s\,\frac{1}{\Gamma}\left(\frac{t}{T}\right)^{s-1} - h'_f\left[1-\left(\frac{t}{T}\right)^{s}\right]^{2} \tag{9.16}$$

whose maximum, reached at the end of the movement, is

$$\Delta h'_{max} = s\,\frac{1}{\Gamma} \tag{9.17}$$

independently of friction (total friction recovery). This maximum depends only on s and Γ. In contrast, the elastic response depends on all three main parameters, namely, ρ, h_f and Γ (or θ) for a given s.

Frictionless flow and $\theta \le 1$. The elastic model gives the Joukowsky head rise $\Delta h'_{max} = 2\rho$, independently of s. The relative error between models is then

$$\varepsilon = \frac{h'^{El}_{max} - h'^{Rig}_{max}}{h'^{El}_{max}} = \frac{2\rho - \dfrac{s}{\Gamma}}{2\rho} = 1 - \frac{s}{2\theta} \tag{9.18}$$

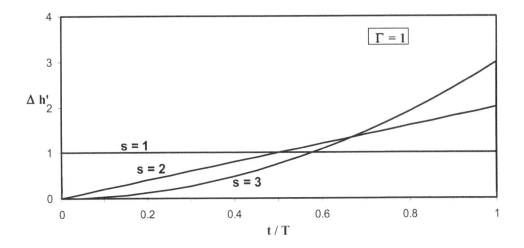

Fig. 9.4 Typical head profiles at valve for s = 1, 2 and 3 (f=0 and rigid model)

showing that for selected s the error is only a function of θ. Numerical results show that this statement remains valid regardless of θ for frictionless flow, provided the discharge law is smooth (s $\geq$ 1).

Frictionless flow and $\theta > 1$. For linear flow reduction (s = 1) the maximum head predicted by the elastic model occurs at time 2L/a and is given by $\Delta h'_{max} = 2/\Gamma$. Because the rigid model gives $\Delta h'_{max} = 1/\Gamma$, the discrepancy is 50%, independently of θ. This result is coherent with eq. 9.18 for $\theta = 1$. For linear discharge variation, it can be easily derived that at t = L/a the maximum pressure head rise is the same for both rigid and elastic models. However, the velocity distribution for the elastic model is linear along the pipe. This is inconsistent with the mass oscillation and is a result of the initial discontinuity in the boundary slope. This lack of smoothness shows that in certain cases compressibility is absolutely essential regardless of the variation of speed. If the initial or boundary conditions or their derivatives are discontinuous, the discontinuities propagate along the characteristics and the solutions are discontinuous in the same sense.

Rigid model shows abrupt change in pressure head at the points of discontinuous derivatives of boundary condition. For s = 1, it occurs at t = 0 and t = T and for s > 1, it occurs at t = T. For rigid model, Fig. 9.4 shows the pressure head variations at valve end for s = 1,2 and 3.

For 1< s <2, the elastic model gives higher pressure values than the rigid one and the maximum head occurs before the end of the movement. In contrast, for s >2 the elastic responses are smaller than those predicted by rigid theory and the maximum occurs at the end of the movement. For parabolic variation (s = 2), if T = nT$_e$, in which n is an integer, the maximum relative pressure head rise coincides with the rigid model result and is equal to $2/\Gamma$. Fig. 9.5 shows maximum head variations for $s = 1, 2$ and 3 for both rigid and elastic model.

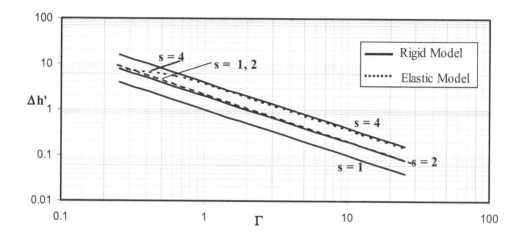

Fig. 9.5 Maximum head rise at valve versus Γ for different values of s ($f = 0$)

Frictional flow. In the presence of friction, the applicability of rigid model, for given s, depends not only on θ but also on ρ and h_f.

For $1 < s < 2$, for which, neglecting friction, the elastic model predicts larger maximum heads, pipe wall friction tends to reduce these magnitudes and consequently the relative error between models. From eq. 9.16, it is clear that any substantial reduction caused by friction may be expected to occur during the initial stage of the closure, when velocity is still important. On the other hand, the elastic model gives the maximum head in the early stage of closure snd so there is only partial head recovery. Moreover, wave reflection effects are also beneficial in reducing transient oscillation for these values of s. For $s = 2$, Fig. 9.6 shows the boundaries of the practical rigid zone (region in which $\varepsilon \leq 2\%$) for some values of $h'_f = 0$, 0.1 and 0.8 as a function of ρ and Γ. It is observed that, as h'_f increases, the practical rigid zone has a tendency to increase. Hence, it is both conservative and practical to use the frictionless case to establish the boundaries of the practical rigid zone.

In contrast, if $s > 2$ the wave effects will tend to increase the transient maximum head and this difference appears to increase as h'_f increases and/or ρ decreases.

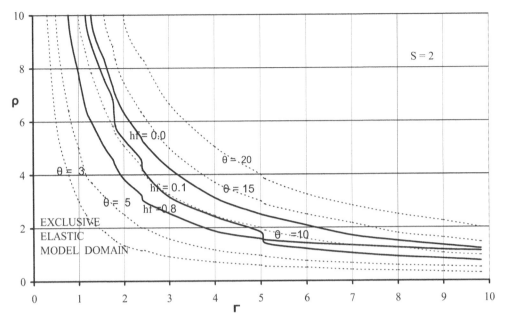

Fig. 9.6 Chart showing practical limits of application ($\varepsilon \leq 2\%$) of rigid model for different h'_f values (s = 2)

The following summarizes the above:

1. In the absence of friction and when a sufficiently smooth prescribed flow rate variation is prescribed at the boundary, the applicability of rigid model only depends on θ.
2. The shape of the prescribed q(t) boundary condition plays an important role in the range of application of the rigid model. Discontinuity in gradients of the prescribed discharge law causes more error.
3. For frictional case, the applicability of the rigid model depends on ρ and h'_f in addition to θ and solutions for small or large values of s is different. For small rates of flow change (s <2), it is conservative to use the frictionless case for the determination of boundaries of the practical rigid zone. However, for higher values of s, the friction effects increases the magnitude of the maximum head and the difference between elastic and rigid results become more pronounced.

Second discharge law

The discharge law derived in chapter 7 for optimal control of pressure surges is considered as second discharge law for deriving the applicability of rigid model. The discharge law from an initial flow rate q_0 to a final value of zero is given by

$$q = \frac{q_o}{2} - \frac{q_o}{2} \frac{\left\{ \cosh \gamma t' - \cosh \gamma (1 - t') + \left(\frac{1}{2} - t' \right) \gamma \sinh \gamma \right\}}{\left(\cosh \gamma - 1 - \frac{\gamma}{2} \sinh \gamma \right)} \tag{9.19}$$

This discharge law guarantees continuous gradients at t = 0 and t = T and the shape varies with the value of γ. For different values of γ (=1, 9 and 25), Fig. 9.7 shows typical discharge profiles.

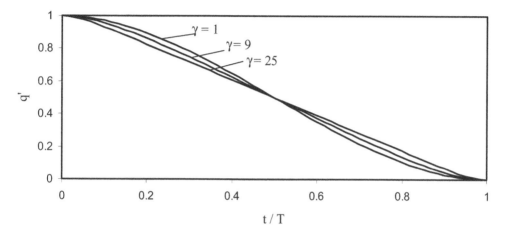

Fig. 9.7 Typical second discharge law profiles for different values of γ

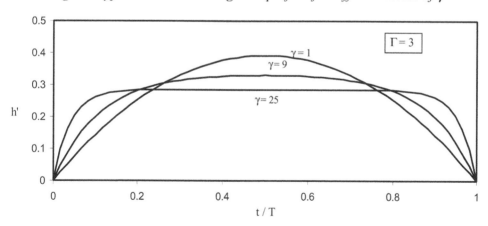

Fig. 9.8 Typical head profiles at valve for different values of γ (f = 0 and rigid model)

Frictionless flow. The rigid model gives maximum head at t = 0.5T and the maximum value depends on Γ and γ. The maximum relative head is given by

$$\Delta h'_{max} = \frac{1}{2\Gamma} \frac{\gamma (2 \sinh(0.5\gamma) - \sinh \gamma)}{(\cosh \gamma - 1 - 0.5\gamma \sinh \gamma)} \tag{9.20}$$

and as Γ and/or γ increases, the maximum relative head decreases. The percentage reduction in maximum head as γ increases is independent of Γ. This reduction is 15.8% for $\gamma = 9$ as compared to $\gamma = 1$. This figure is 27.2% and 30.2% for $\gamma = 25$ and 50 respectively. Fig. 9.8 shows typical head profiles at valve for different values of γ (=1, 9 and 25) Fig. 9.9 shows variation of $\Delta h'_{max}$ with Γ for different values of γ. As it is clear from Fig. 9.8, as γ increases though $\Delta h'_{max}$ decreases, the rate of change of pressure head at the beginning and end of valve movement increases.

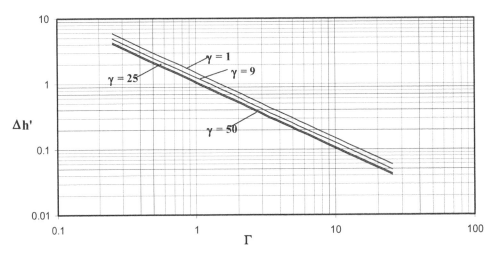

Fig. 9.9 Maximum head at valve versus Γ for different values of γ ($f = 0$ and rigid model)

The maximum relative head in the case of elastic model also depends on ρ except Γ and γ. For a particular value of γ and ρ, as Γ increases the relative error between the models decreases. Obviously, with the increase in the value of ρ, the error reduces. Figs. 9.10 and 9.11 show relative maximum head variation with Γ for different values of ρ for $\gamma = 1$ and 9 respectively.

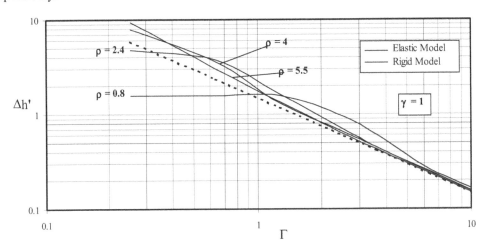

Fig. 9.10 Maximum head at valve versus Γ for different values of ρ ($f = 0$ and $\gamma = 1$)

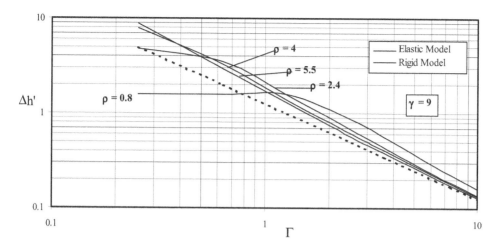

Fig. 9.11 Maximum head at valve versus Γ for different values of ρ (f = 0 and γ = 9)

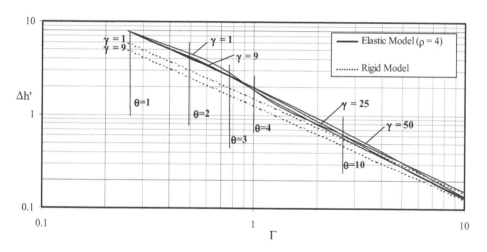

Fig. 9.12 Maximum head at valve versus Γ for different values of γ (f = 0 and ρ = 4)

In contrast to the rigid model where as γ increases, maximum pressure head decreases, elastic model shows different trend. With the increase in the value of γ, there is a slight decrease in the maximum head in the beginning but it generally increases. This is because as γ increases the discharge function has more relative weight on the decrease of head as compared to rate of change of head. However, due to increased rate of change of head, elastic model produces more oscillations and thus increasing the magnitude of the maximum head. Fig. 9.12 shows maximum head variation with Γ for different values of γ for ρ = 4.

Fig. 9.13 shows some typical head profiles at valve for different values of Γ, θ and γ. It is evident that head profile in case of elastic model oscillates around head profile of rigid model and oscillations are more as γ increases.

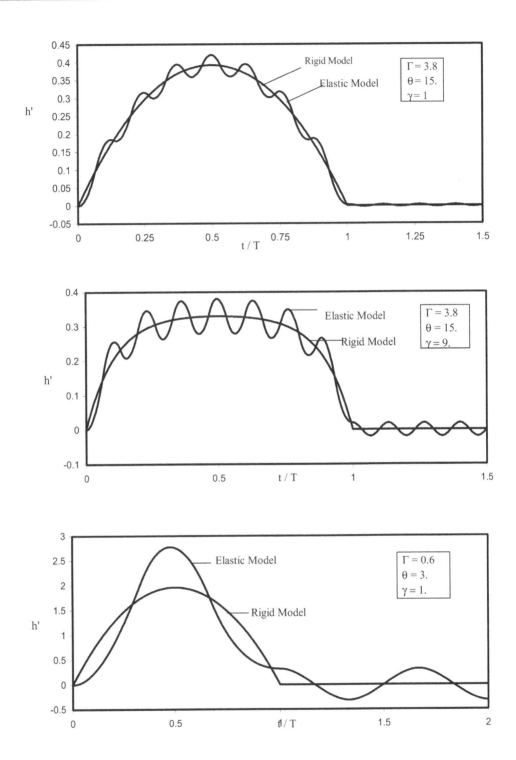

Fig. 9.13 Typical head profiles at valve for different values of Γ, θ and γ. (f = 0)

Frictional flow. When friction is taken into account the relative error between the models reduces. This reduction is more pronounced for $\theta \geq 5$ and as h'_f increases. For $\theta < 5$, there is almost no significant reduction in relative maximum heads. Figs. 9.14 and 9.15 show relative error in maximum heads between the two models for different values of Γ for $\rho = 5.5$ and 4 respectively. For $\rho = 5.5$, the error is less than 10% if $\theta > 7$ for $h'_f = 0.5$ and if $\theta > 10$ for $h'_f = 0.2$ or 0.

Fig. 9.16 shows typical head profiles at valve for $h'_f = 0.2$. As in the case of frictionless case, head profile of elastic model oscillates along the profile of rigid model. Obviously, amplitude of oscillation reduces due to friction.

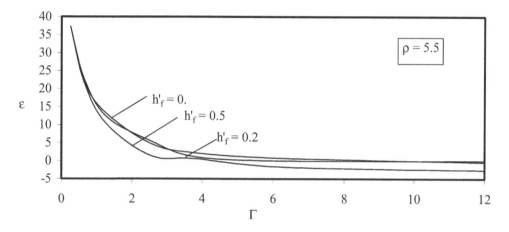

Fig. 9.14 Relative error in maximum head between elastic and rigid models versus Γ for different values of friction ($\rho = 5.5$)

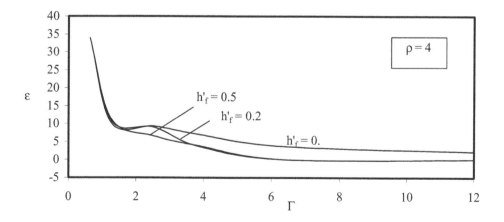

Fig. 9.15 Relative error in maximum head between elastic and rigid models versus Γ for different values of friction $\rho = 4$)

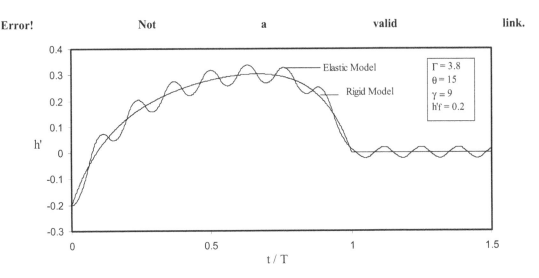

Fig. 9.16 Typical head profiles at valve with $h'_f = 0.2$

This summarizes the above:

1. Second discharge law represented by eq.9.19 is more smooth than the first discharge law in the sense that it possesses continuous gradients both at time t = 0 and t = T. The effect of this is the qualitative change in the elastic response. The elastic response, in fact, follows the rigid response with some oscillations (see Figs. 9.13 and 9.16). The amplitude of the oscillation depends on ρ, Γ, h'_f and γ. The relative error in maximum head obtained from two models is, in fact, due to these oscillations only. As ρ, Γ and h'_f increase, the relative error reduces.

2. As γ increases, the rigid model shows decrease in maximum head. How ever, the rate of change of head change increases. The effect of this is reflected in elastic response wherein the amplitude of the oscillations increases and, thus, higher relative error between the models. In absolute terms also, as γ increases, the maximum head due to elastic response decreases in the beginning but it generally increases for higher values of γ.

From the above parametric analysis for the maximum head response from rigid and elastic models, it can be concluded that the use of simplified model (rigid model) for the analysis and control of hydraulic systems depends not only on the system's inertia, elasticity and friction but also on the type and shape of forcing function i.e. boundary condition. Though techniques are well established for the analysis of a complex hydraulic network using elastic model, the problem of design of valve operations for the specified transients in a complex system is not straightforward. If system's characteristics and operational requirements permit the use of rigid model with acceptable error limits, valve operations designed using rigid model can well be used.

In the next section, some examples of hydraulic systems are considered to illustrate the use of rigid model for the analysis and control of transients.

9.3   Control of Transients Using Rigid Model Valve Operations

In this section, some hydraulic systems are considered for the problem of control of transients. Design of valve operations are made using rigid model using the principles developed in chapter 7. The designed valve operations are then used to get elastic response of the system and, thus, to compare the results of the two model.

For calculating of elastic response, explicit method using method of characteristics is used.

Example I

As a first case, the single reservoir-pipe-valve system shown in Fig. 9.1 is considered. Though the procedures for control of transients such as valve stroking in specified time or specified maximum limiting head, for this system are well established, the example is the most simple hydraulic system and has been considered by many investigators.. The valve operations designed using rigid model are used for to check the elastic response of the system.

Consider a pipe of 1000 m length, 1.284 m diameter and having a coefficient of friction as 0.016. Reservoir head is 50 m. It is desired to change the flow from an initial value of 5 m³/s to zero in 30 seconds. Value of γ is taken as 1. Valve operations obtained using rigid model are shown in Fig. 9.17 (a). These valve operations are used to obtain the elastic response of the system and head at valve is drawn in Fig. 9.17(b) for both models. Wave speed for elastic model is 1000 m/s.

In the example, $\rho = 3.94$, $\theta = 15$ and h'_f is 0.2. Results show that the maximum head at valve from rigid model is 67.7m. If designed valve operations are used to get the elastic response, the maximum head obtained is 57.6m. The elastic response follow the same profile as obtained in rigid model but some residual transients remains.

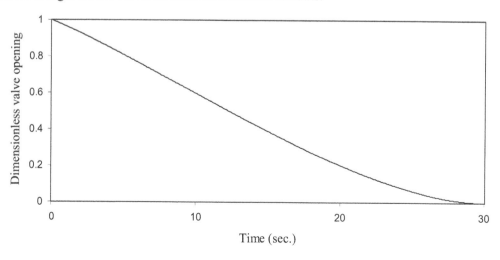

Time (sec.)

(a)

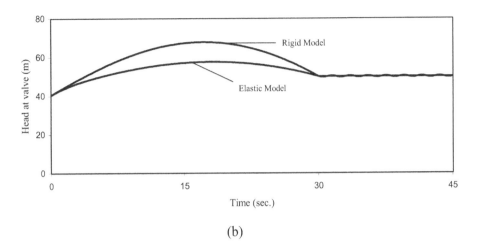

(b)

Fig. 9.17 Results of exampleI: (a) Dimensionless valve openings using rigid model; (b) Head at valve from elastic (using valve operations of (a)) and rigid models

Example II

A branched network shown in Fig. 9.18 is considered as second example. Length of pipe number 1, 2 and 3 are 1000m, 1500m and 1000m respectively. All pipes have D = 1.2m and f = 0.016. Reservoir head is 50m. Initial discharge in pipe number 1, 2 and 3 are $4m^3$/s, $3m^3$/s and $7m^3$/s. It is desired to reduce the flow in pipes 1 and 2 to $1m^3$/s in 30 seconds. $\gamma = 1$ and a = 1000m/s is taken.

From rigid model, valve operations obtained to achieve the above objective are shown in Fig. 9.19(a) and the head variations at point A, B and C are shown in Fig. 9.19(b). When these valve operations are used for transient analysis using elastic model, maximum value of head increases slightly (see Fig. 9.20). However, valve operations provide smooth transients.

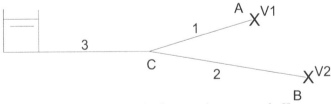

Fig. 9.18 A branched network as example II

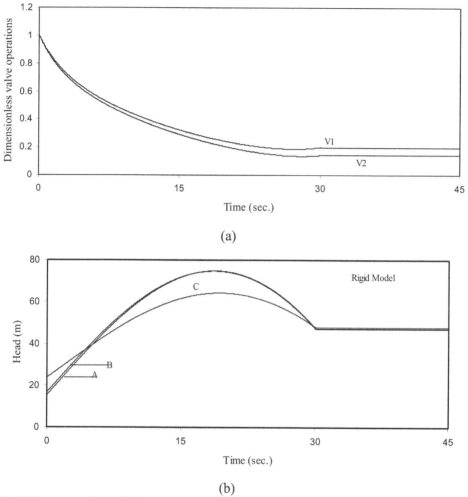

(a)

(b)

Fig. 9.19 Example II: (a) Valve operations obtained from rigid model; (b) Head variations from rigid model

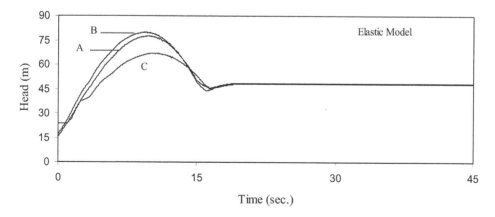

Fig. 9.20 Example II: head variations at nodes from elastic model using valve operations of Fig. 9.19 (a)

Example III

Pipe network shown in Fig. 7.6 in chapter 7 is considered as next example. The network is redrawn here in Fig. 9.21. Applications of the procedure developed for optimal control of pressure surges in chapter 7 were shown by considering three cases. Case I was the full closure of valves from an initial steady state. The designed valve operations for this case are shown in Fig. 7.7. Network's characteristics and two steady states, initial and final, are given in Table 7.2 and 7.3 respectively.

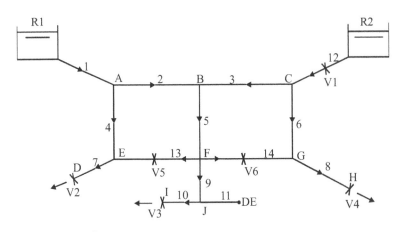

R : Reservoir
V : Valve
DE : Dead End

Fig. 9.21 A pipe network considered as example III

To show the applicability of rigid model results, the valve operations designed for case I are applied and network is analysed using elastic model. For this purpose, a wave speed of 1000 m/s is considered in each pipe. Fig. 9.22 shows the head variations at nodes A, B, C, E, F and G obtained using elastic model and rigid model.

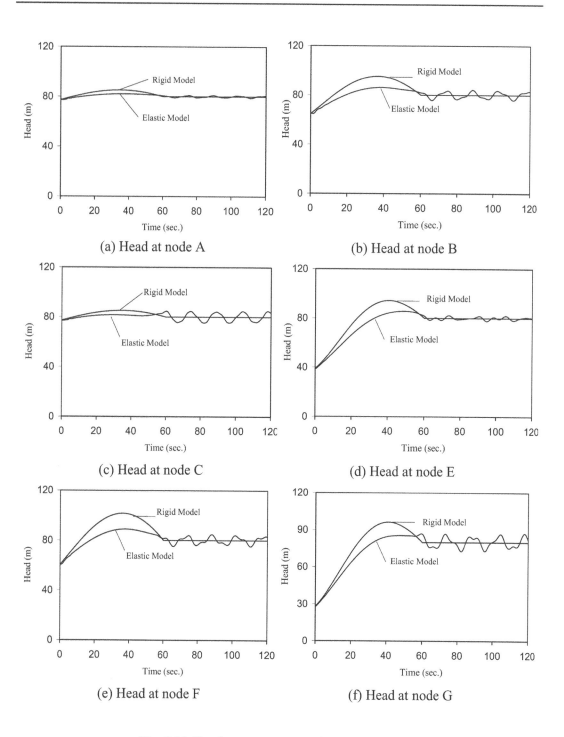

Fig. 9.22 Head variations at nodes of example III

9.4   Control of Transients Using Elastic Model

For the control of transients using elastic model , procedures of valve stroking are well developed for some simple networks (Streeter 1963, 1967; Propson 1970). Valve stroking aims at transfer of the system from an initial steady state to a final steady state within specified time or specified maximum head without any residual transients. The flow condition during the process can be described by pressure surges as no flow reversal takes place.

Consider the case of a single reservoir-pipe-valve system as shown in Fig. 9.1. If valve stroking is to be achieved within time T, a discharge variation from an initial value to a final value is applied at reservoir end for time equal to T-2l/a and head and discharge variations at valve end are computed. Linear variation of flow at reservoir end produces minimum head rise at valve end.

Solution of equations 9.1 and 9.2 using characteristic method and explicit finite difference scheme can be written in terms of head and discharge at any point i at time t

$$H_i = H_{i-1} + Bq_{i-1} - Bq_i - rq_{i-1}|q_{i-1}|$$
(9.21)

$$H_i = H_{i+1} - Bq_{i+1} + Bq_i + rq_{i+1}|q_{i+1}|$$
(9.22)

where head and discharge at points i-1 and i+1 are at time t-dt. Values of dx and dt are selected so that dx/dt = a. B is equal to a/gA.

From equations 9.21 and 9.22, it is clear that minimisation of head requires minimisation of change in discharge between two time steps. In other words, dq/dt shold be minimum between two time steps. This gives a linear change in flow to produce minimum head change.

Consider the second discharge law given by eq. 9.19 and apply it at reservoir end. For l = 1000 m, D = 1.0 m, f = 0., H_R = 50 m, T = 30 seconds, q_o = 5 m^3/s and q_T = 0, the head variations at valve end for γ = 1 and 9 are shown in Fig. 9.23.

As the value of γ increases, the head change reduces. It is clear from the derivation of this discharge law carried out in chapter 7 that as the value of γ increases, more weightage is given to minimisation of dq/dt. The head change at valve end in this case (γ = 9) is equal to that obtained by linear variation of discharge at reservoir end.

Figs. 9.24 and 9.25 shows discharge variations and dimensionless valve coefficients for these two cases; γ = 1 and 9.

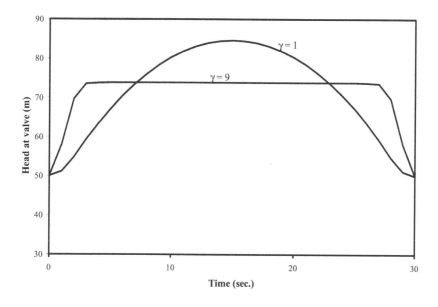

Fig. 9.23 Head variation at valve end

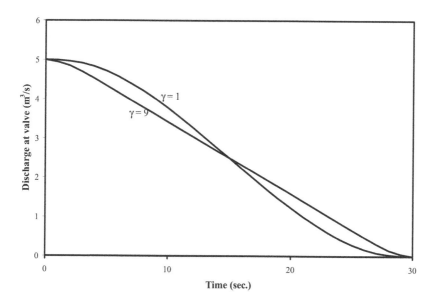

Fig. 9.24 Discharge variations at valve end

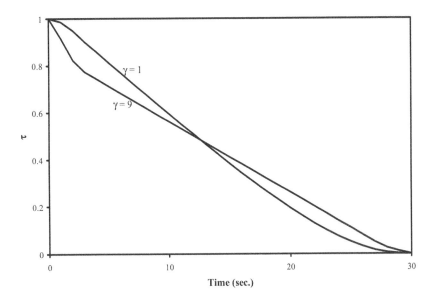

Fig. 9.25 Dimensionless valve coefficients

For the specified maximum head, the value of γ can be obtained using the procedure developed in chapter 7 and the discharge law given by eq, 9.19 may be used at reservoir end.

Procedures of valve stroking have been extended to tree like networks with valves at the end of each branch. These procedures can further be extended to looped networks also. However, as per network controllability condition, every loop should posses a valve. Moreover, the location of valves in every chord should be at the end in the direction of flow. Fig. 9.26 shows two simple networks with the appropriate position of valves for valve stroking.

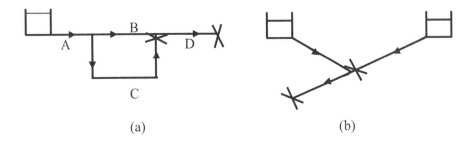

(a)   (b)

Fig. 9.26 Pipe networks having valves in every chord

Consider pipe network as shown in Fig. 9.26 (a). With a specified discharge at reservoir, pipe A can be valve stroked. With a distribution of discharge at the junction of pipe A, B and C, pipes B and C can be valve stroked. The difference in head at the downstream end of pipe B and C will give the head loss in valve located in pipe C. Finally knowing head at the junction of pipe B and discharge at upstream of pipe D, pipe D can be valve stroked.

9.4.1   Control of transients with defined residual transients

Valve stroking procedures produce high head at valves and valves have to act against high head. If the requirement is to obtain a flow condition without residual transients in the system at the end of valve operations, procedures of valve stroking are necessary. However, in many cases of engineering applications, some residual transients are not considered objectionable. In such cases, head change in the system due to valve operations can further be reduced.

The objective of the prolem in this case is to design the valve operation rules to transfer the system from initial to final steady state for the given maximum head change or max./min. head in the system and the given valve operation time, T. In this case, though valve operations cease at time T and thus valve openings reach final steady state values at time T, but head and discharge do not reach their final steady state values at time T.

Consider the case of a single reservoir-pipe-valve system as shown in Fig. 9.1 and consider frictionless case. The x-t diagram of this system representing flow conditions is shown in Fig. 9.27.

In the figure, point D is at valve end at time T i.e. at the time when valve operations ceases. At this point, let the residual transients i.e. head and discharge be H_{res} and qres. The values of residual discharge and head at point D are related by the following equation for the valve:

$$\frac{q_{res}}{q_o} = \tau_f \sqrt{\frac{H_{res}}{H_o}} \tag{9.23}$$

where

$$\tau_f = \frac{q_f}{q_o} \sqrt{\frac{H_o}{H_f}} \tag{9.24}$$

q_o, q_f, q_{res} are initial steady state discharge, final steady state discharge and discharge at time T through the valve. Similarly, H_o, H_f, and H_{res} are initial steady state head loss, final steady state head loss and head loss at time T throgh the valve. In eqs. 9.23 and 9.24 head loss through the valve is written in terms of head at upstrea end of valve after considering, downstream head at valve as zero. τ_f is dimensionless valve coefficient for final steady state .

The above equation guarantees that valve reaches its final steady state opening at time T, if discharge and head at valve at time T reach qres and Hres..

If qres and Hres are discharge and head at valve at time T i.e. at point D in the x-t diagram, discharge at point C can be calculated using C^+ characteristics from backward calculations. Let discharge at point C i.e. at reservoir end at time T- L/a be q_c. The C^+ characteristic equation from point C to point D becomes

$$H_{res} = H_R - B * (q_{res} - q_c) \tag{9.25}$$

where $B = a/gA$ and H_R is reservoir head.

For the case of valve stroking in time T, the maximum head rise in case of valve closure is given by (Wylie and Streeter, 1993),

$$H_{max} = H_R + \frac{B}{\left(T - \frac{2L}{a}\right)} \frac{L}{a}(q_o - q_f) \qquad (9.26)$$

where L is length of pipe and a is wave velocity. In valve stroking case, discharge at point C and D is q_f.

In the present problem, if discharge at point C is q_c, the maximum head rise in the pipe is related to q_c by the following equation:

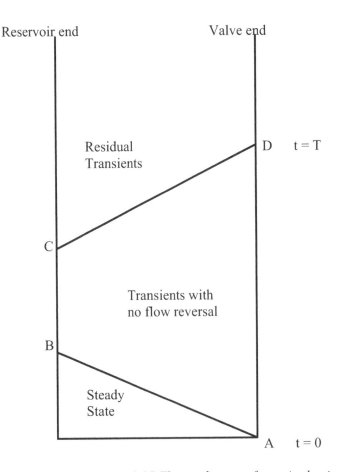

Fig. 9.27 The x-t diagram for a single pipe

$$H_{max} = H_R + \frac{B}{\left(T - \dfrac{2L}{a}\right)} \frac{L}{a}(q_o - q_c)$$
(9.27)

The maximum head rise in this case is considered to occur at time from t=2L/a to t=T-2L/a as obtained in valve stroking case.

By defining the the maximum head in the pipe, value of discharge at point C, q_c, can be obtained from eq. 9.27. Knowing the value of q_c, H_{res} and q_{res} can be obtained from eqs. 9.23 and 9.25. Value of τ_f in eq.9.23 can be computed from eq. 9.24. After calculating the value of H_{res} at point D, a smooth head variation from point A to D can be applied with known value of H_{max} which is considered to occur from t=2L/a to t=T-2L/a.

Knowing the values of head at upstream and downstream end of pipe, values of discharges at different points can be calculated in the region enclosed by points ABCD from characteristic equations. This procedure guarantees valve opening to reach its final steady state stage at time t = T, though flow will not reach its final steady state stage at that moment. Above line CD, simple analysis of transients is made, where the upstream boundary condition is H = H_R and the downstream boundary condition is given by valve equation with constant τ (= τ_f).

In the above procedure, if a in the head variation from point A to D, head is not kept constant from time t = 2L/a to t = T-2L/a and a dfferent profile is assigned, the relationship between H_{max} and q_c shall be obtained first from charateristic equations in the region ABCD.

To illustrate the above procedure, consider a single reservoir-pipe-valve system with l = 2000 m, a = 1000 m/s, A = 10.0 m^2, f = 0, H_R = 50 m, T = 4L/a seconds, q_o = 5 m^3/s, q_f = 2 m^3/s and H_{max} = 160 m.

The maximum head at valve end obtained by valve stroking procedure in this case will be 200m Fig. 9.28 shows head at valve end. Dotted line in the same figure shows the case of valve stroking. The discharge variation is shown in Fig. 9.29.

In this example, q_c works out to be equal to 2.8 m^3/s and values of H_{res} and q_{res} are 78.8m and 2.51 m$_3^3$/s. τ at t = T is 0.4. A smooth head variation is assigned at valve end from time t = 0 to t = T with H_{max} = 160 m occuring at t = 2L/a seconds.

In this example, residual discharge works out to be about more than q_f by 25% of q_f In other words, to reduce the head change of 40 m (which is about 26.7 % of maximum head change obtained in case of valve stroking), a discharge variation of 25% of q_f is obtained at time T.

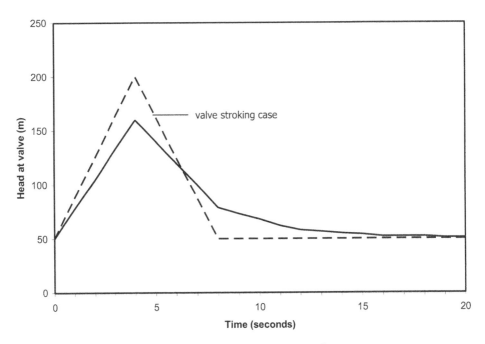

Fig. 9.28 Head variation at valve

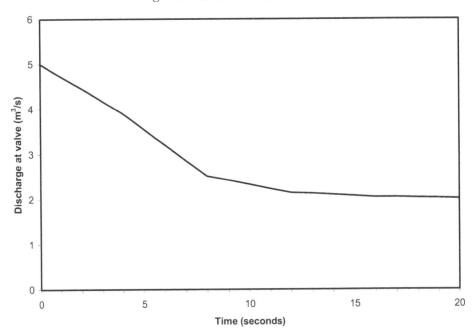

Fig. 9.29 Discharge variation at valve

Consider the same example with T = 6L/a and H_{max} = 95m. The valve stroking maximum head at valve will be 125m in this case. Head and discharge at valve are plotted in Figs. 9.30 and 9.31.

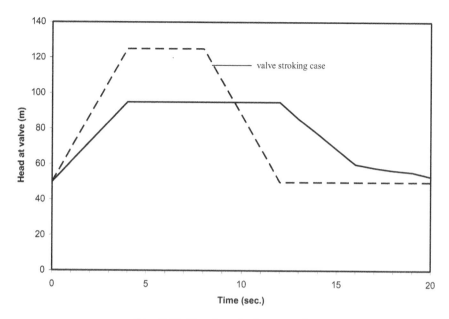

Fig. 9.30 Head variation at valve

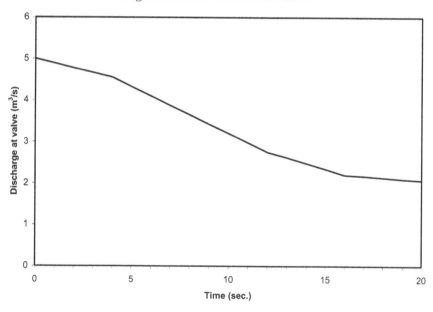

Fig. 9.31 Discharge variation at valve

In the above example, q_c works out to be 3.2 m³/s and H_{res} & q_{res} works out to be 95 m and 2.75 m³/s respectively. In the example H_{max} and H_{res} are equal. If in the above case H_{max} is reduced below 95m, the vaue of H_{res} becomes greater than 95m. If it is desired to keep the maximum head rise in the system limited to some value throughout the time of transients including the time of residual transients, H_{max} can be reduced to only some limited value. This value can be obtained by equating $H_{max} = H_{res}$ in equations 9.23, 9.25 and 9.27.

9.5   Concluding Remarks

The first aim of this chapter was to investigate the applicability of the rigd model in analysis and control of transients and to demonstrate the use of rigid model results. Though, investigations have been carried out in the past to define the boundaries between rigid and elastic models, but most of the studies considered linear valve closure and frictionless flow only. In transient design or transient synthesis approach, head and/or discharge variations are specified and valve operations are designed. This asks for the new investigations for determining the influence of prescribed discharge and/or head variations in deciding the range of application of rigid model.

In the chapter, the applicability of the rigid model has been investigated for a simple reservoir-pipe-valve system keeping in view the synthesis approach to the transients control problems. Influence of prescribed discharge laws is investigated for both frictionless and frictional flow. Results show that shape of the prescribed discharge law influences the boundary defining the applicability of the rigid model. It is found if discharge laws possess discontinuous gradients, great discrepancies in the rigid model results are obtained. For the prescribed discharge law, as obtained by using the procedures of optimal control of pressure surges outlined in chapter 7, results show that elastic response follow the same profile as of rigid one with some oscillations. The results show that the elastic response is in good agreement with rigid one, both qualitatively and quantitatively, for a certain range of parameters defining opearational and system's characteristics. The discrepancy between the two models results depend not only on the system's characteristics and operational parameters but also on the shape of the prescribed discharge law.

The second aim of this chapter was to investigate the elastic response using the results obtained from rigid model. For this, valve operations designed for the optimal control of transients using rigid model are applied to get the elastic response of the hydraulic systems. Results are promising in the sense that elastic response follow the same rodile as of rigid one with some error in maximum/minimum heads. Moreover, the transients resulting from elastic model are also smooth, though some residual transients exists. It can be concluded that for hydraulic systems in which very fast change in flow conditions is not required, procedures of optimal control of transients based on rigid model can be used to design the valve operations.

Procedures for valve stroking for the control of transients using elastic model have already been developed. However, an investiation for further reducing the maximum head change by allowing residual transients has been carried out for a simple system. In many engineering applications, some residual transients are not considered objectionable. The procedure developed in the chapter outlines the design of transients with residual transients. These procedures are to be extended to more complex system.

10 | General Conclusions and Recommendations

Though major results and conclusions of this study are given at the end of every chapter, this chapter presents first some general conclusions of the present study. Then some extensions are briefly addressed as potential study topics.

10.1 General Conclusions

Conclusions concerning synthesis approach to network analysis and control

1. Growing demand and complexity of hydraulic networks has led to the constant research efforts by engineering community for efficient design and management of these systems. Area of analysis and control of networks has been intensively explored by researchers, however, expected results have not been achieved. Reasons for this may be attributed to the physical and topological complexity of these networks and development of non-coherent approaches to solve different types of network problems associated with steady and unsteady flows. In fact, scientific community has viewed different network problems from different perspectives.

   Problems of network analysis, design and operation can be dealt by two approaches; analysis approach and synthesis approach. Analysis approach uses trial and error procedure to evaluate design variables. Most of the current codes such as EPANET, WATSYS, WANDA etc. are based on analysis approach. Synthesis approach has not been fully developed, though some efforts of the researchers in this direction can be seen in the literature.

2. Synthesis approach to network analysis and control aims at direct evaluation of design variables for the specified hydraulic behaviour of the system. Using this approach, mathematical formulation of network problems is not straightforward because problem solvability depends on the manner in which design variables and specifications are topologically allocated.

   This study provides a general network synthesis framework for approaching network design and control problems from synthesis perspective. As per the developed framework, the first step for solving network problems is to categorise systematically all boundary conditions and to develop condition, for network solvability. Then mathematical formulation and development of solution algorithm follows. This framework is general in the sense that different types of network problems related to both steady and unsteady flows can be approached from the same perspective.

The study shows that synthesis approach provides an efficient and simple platform not only for solving practical engineering problems of design and operation hydraulic network but also for understanding the hydraulic behaviour of these systems. In the study, applications of this approach are shown in the area of system design, transient design and transient system component design.

Conclusions concerning network modelling and network solvability rules

1. Network modelling comprises of three steps: component modeling, problem formulation and development of algorithm for solution. These three steps are equally important for the development of an accurate and efficient network model. Network component modelling should be generalized and should take care of all possible components encountered in practice. A well-posed problem and its mathematical formulation is the key to obtain an accurate and unique solution. If problem is not well posed, multiple or no solution may be obtained. It is necessary for an meaningful network model that it satisfies the conditions for the existence and uniqueness of a solution. Moreover, mathematical formulation should be such that dependent and independent variable are distinguished clearly and relationships between them are shown explicitly. Algorithm for the solution should provide fast and accurate results.

This study has outlined the importance of an effective network modeling. In the study, network models have been developed for problems related to both steady and unsteady flow. Network models are generalized in the sense that models can handle arbitrary topology and possible boundary conditions encountered in practice.

2. A network consists of different types of components, constraints and demands. All these represent different kind of boundary conditions, which are mutually interacting. This study has systematically categorized these different boundary conditions into six types of elements. These six types of elements are classified based on known/unknown values of head loss and discharge variables. Study shows that this division of different boundary conditions into six types of elements is comprehensive in the sense that a variety of possible practical problems can be modelled by these six types of elements.

In this study, these six types of elements have been given name as element type 1, 2, 3, 4, 5 and 6. Such names are easy to remember and have the advantage of going well with computer implementation of the problem.

3. Synthesis approach to network analysis, design and control aims at design of system variables for the specified hydraulic behaviour of the system. For a well-posed problem and its unique solution, these specifications and design variables cannot follow arbitrary topology.

The study has developed conditions for a well-posed problem and network solvability in terms of necessary and sufficient conditions for the existence and uniqueness of a solution. The seven necessary conditions and one necessary & sufficient condition, which are defined by eight theorems, must be satisfied by a network model of any problem.

These conditions put restriction on topological allocation of specifications and design variables. Briefly, these conditions can be stated as : one known head, no loop 1 & 3 no loop 1 & 4, no cut 2 & 3, no cut 2 & 4, ne3 = ne4, existence of T_{13a} and T_{14a} and non-singular matrix $M (= r_b + L_{ba} r_a L'_{ba}^T)$

Study shows that these conditions which have been developed using graph theory, are helpful in demarcating different network properties. One of the major results of this study is the development of the necessary condition about existence of two spanning trees, T_{13a} and T_{14a}. Its usefulness has been shown in mathematical formulation and development of algorithm for solution of problems concerning both steady and unsteady flows. Algorithm for generating these spanning trees has been developed in the study, which is easy to implement on computer.

It is to mention here that one of the main reasons for not much development in the past in the area of synthesis approach to network analysis and control, was the absence of a generalized network model and insufficient network solvability conditions. With the development of a generalized network model and solvability conditions in the present study, the usefulness and potential of synthesis approach is shown to solve a variety of engineering problems associated with network analysis, design and control.

Conclusions concerning network analysis, design and control

1. The current study shows that synthesis approach to network analysis and control has a vast potential in the area of system design, transient design, transient system component design, operational optimization etc. In the study, different kinds of network problems concerning steady and unsteady flows have been classified systematically and have been treated in different chapters. Different problems have been viewed from a common platform, which is based on network synthesis framework. Applications of the developed procedures are shown in the area of network analysis, design, control and calibration.

2. In the study, problems concerning steady flow have been divided into two types: analysis problem and analysis design problem. Using the result of existence of spanning trees, T_{13a} and T_{14a}, the problem formulation is well structured. Mathematical formulation using this condition provides automatic separation of independent and dependent variables. Moreover, a minimum set of independent variables is obtained automatically. The other advantage of the algorithm is that it provides explicit calculation of system parameters, as these parameters are not part of independent variables.

    The developed algorithm with the above stated advantages among others is a powerful tool not only for efficient solution of engineering problems, but also for learning and understanding the interrelationships between different variables and hydraulic behaviour of system in general

The non-linear programming formulation of steady state problems is carried out using optimisation approach. Formulation has developed objective functions, which are based on content and co-content of the network. These objective functions are convex in nature and solution of network problem is obtained by minimisation of these functions. Development of this method provides an algorithm for network solution which guarantees convergence to the sought solution.

Study shows a variety of applications of analysis and analysis-design problems in network design, operation and calibration.

3. Conventional approach to transient control and transient system component design uses trial and error procedure of analysis approach. The present study has developed procedures for control of transients and system component design using synthesis approach. The procedure can be termed as transient design. Transient design and transient system component design aims at design of boundary conditions and system components for the specified transients in the system. Procedures have been developed for both pressure surges i.e. transients analysed using rigid model and transients analysed using elastic model.

Control of pressure surges using valve operations aims at design of valve operations for the specified transients. Problem can be defined as design of valve operations for transferring a network from an initial steady state to a final steady state within a specified time. At the end of time T, valve operation ceases. Presence of residual transients depends on the network topology. In the study, a criterion for network controllability has been developed. As per this criterion, for the full control of pressure surges, number of valves in the network should be equal to number of loops in the network and their location should be in the chords of the network. Networks following the above criteria can be transferred to another steady state without any residual transients.

Based on number of specifications and design variables along with the topology of the network, the study has divided systematically control problems into four types. These four types of control problems are determined full control problem, determined partial control problem, underdetermined full control problem and underdetermined partial control problem.

Determined problems use only a numerical technique for the solution of a set of ordinary differential equations. However, formulation of these problems is not straightforward. Network model must follow necessary and sufficient conditions for the existence and uniqueness of a solution. The current study provides algorithms for the solution of these problems. Problem formulation utilises network topological properties in the sense that it utilises the presence of two spanning trees T_{13a} and T_{14a}. Algorithm provides automatic separation of independent and dependent variables and provides a minimum set of ordinary differential equations describing the hydraulic behaviour of the system. One of the main advantages of the algorithm is that valve operations are not part of independent variables and these are calculated explicitly.

Underdetermined problems of pressure surges control utilise minimisation of an objective function. In the study, a quadratic objective function is developed which aims at minimisation of change in head and time rate of change of head in the system. Procedure is developed using the principles of calculus of variations, which provides analytical solution of the problem.

Study shows a variety of applications of these developed procedures in the area of transient control, operation optimisation, on-line control and system design.

4. This study provides a generalised algorithm for transient system component design and shows its application in the design of surge tank and valve operations. Algorithm utilises optimisation techniques and develops procedures for the direct design of these components. Traditional approaches for transient system component design uses trial and error procedures. The present algorithm avoids trial and error procedure and provides design parameters of components directly for the specified transients in the system. These procedures are helpful in system design and optimisation and provides an insight into the relationships between different components in the system.

5. For the control of transients using elastic model, study highlights the importance of using rigid model results. For cases where change in flow is slow, transient control can be done using the results of rigid model. Valve operations obtained using the rigid model can well be applied in such cases to control transients.

The study has extended the valve stroking principles to include looped networks. In looped networks, for valve stroking to be possible, there should be a valve in every loop and its location should be at the extreme end of each chord.

Valve stroking requires valves to act against very high heads. Hence, these procedures are not helpful unless a case specifically requires valve stroking. In fact, in most cases, presence of residual transients is not objectionable. The study has developed a procedure for transient control with the specified residual transients or reduced maximum change in head in the system. Allowing residual transients further reduces the head in the system.

The present study has developed the concept of network synthesis and has shown its application in system design, transient design and transient system component design. Study shows that the developed approach is helpful in solving a variety of engineering problems related to network analysis, design, control and calibration.

10. 2   Recommendations for further study

In this section, some extensions of the current study have been addressed as study topics, which can be pursued in future.

1. Underdetermined steady state problems have not been addressed in the present study. These problems will have more number of type 4 elements than type 3 elements. Obviously, problem requires minimisation of an objective criterion. Most useful criteria is

the minimisation of cost. Though the researchers have intensively dealt cost optimisation problem, solution of the problem using the developed principles and results may simplify the formulation and thus, the efficiency of the algorithm.

2. Procedures for transient system component design requires specifications of transients in the system. Design of component is made which satisfy these specifications. The results of the developed procedure will be accurate if the transients specifications matches well with the actual transients which are obtained when these components are placed. In complex networks, generally, it is difficult to ascertain the complete head and discharge variations before the actual design of component. Though smooth discharge variations may be given as specifications, but it is difficult to design head variations. Procedure for the design of specifications in such cases should be developed.

3. The developed procedure for control of transients using elastic model should be extended to looped networks. Procedure of transient control with residual transients is helpful in reducing the pressures in the system below than that obtained by using the principles of valve stroking. Further extension of the study is required for control of transients in complex networks.

Appendix A | Networks and Theory of Graph

Only the barest properties of a directed graph are introduced here by way of background (Chen, 1990). These properties are used in the mathematical formulation and solution algorithm of different network problems.

In general, a graph G is a set of nodes and elements interconnected at nodes.

Node connectivity of a node is defined as the number of elements connected to a node.

A tree T of a connected graph G is defined as any sub-graph so that it; (i) is connected, (ii) contains all the nodes of graph G and (iii) has no loops. A tree has exactly one path between every pair of nodes. A graph containing 'nn' number of nodes has exactly nn-1 number of elements in its tree.

Co-tree T' of a graph is defined as the complement of tree T, the sub-graph of connected graph G that remains after deleting all the elements of the tree sub-graph. If a graph G has 'ne' number of elements, its co-tree contains exactly ne-nn+1 elements in it. The elements in the tree are called branches and in the co-tree are called chords.

The incidence of the nodes and elements of a graph G may be described algebraically by means of matrix, N, called node incidence matrix. In node incidence matrix, rows correspond to the nodes of G and columns correspond to its elements. Hence, the size of a node incidence matrix is nn × ne. The entries of node incidence matrix, N_{ij} are +1, -1 or 0, as follows:

N_{ij} = 0, if element j is not incident with node i.

N_{ij} = +1, if element j is incident with node i and oriented away from it.

N_{ij} = -1, if element j is incident with node i and oriented towards it.

Fig. A.1 shows a directed graph and its node incidence matrix. Graph has 7 nodes and 10 elements. In the figure, numbers along the node and edges show node and element number respectively.

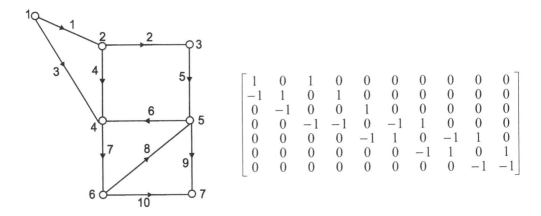

Fig. A.1 A directed graph and its node incidence matrix, N

A node incidence matrix can be subdivided in terms of branches and chords. If all the branches of a tree are grouped together and all the chords are grouped separately, the node incidence matrix can be written as

$$N = [N_t \quad N_c] \tag{A.1}$$

where N_t and N_c are node incidence sub-matrices for a spanning tree elements i.e. branches and co-tree elements i.e. chords respectively.

If the row corresponding to reference node is removed, the size of node incidence matrix becomes nn-1 × ne and size of sub-matrices N_t and N_c are nn-1 × nn-1 and nn-1 × ne-nn+1. Sub-matrix N_t corresponding to the branches of a spanning tree is a non singular matrix.

The addition of a chord to the spanning tree forms a loop. Corresponding to ne-nn+1 number of chords, there are ne-nn+1 number of loops called fundamental loops. The number of fundamental loops, nl, in a graph G are, thus, given by

$$nl = ne - nn + 1 \tag{A.2}$$

The incidence of the loops with the elements of a graph G is described algebraically by means of a matrix L called loop incidence matrix. In loop incidence matrix, rows correspond to loop number and columns correspond to elements of the graph. Hence, size of a loop incidence matrix is nl × ne. The orientation of a loop is taken to be the same as that of the defining chord. The entries of loop incidence matrix, L_{ij}, are +1, -1 or 0, as follows:

$L_{ij} = 0$, if element j and loop I are not incident.
$L_{ij} = +1$, if element j is incident with loop i and their orientations coincide there.

$L_{ij} = -1$, if element j is incident with loop i and their orientations opposed there.

Fig. A.2 shows a directed graph and its loop incidence matrix, L. The graph has 4 loops and element number 3, 5, 7 and 10 are taken as chords as shown by dotted lines.

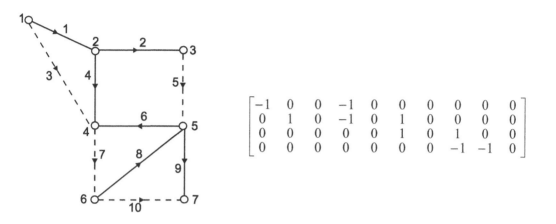

$$\begin{bmatrix} -1 & 0 & 0 & -1 & 0 & 0 & 0 & 0 & 0 & 0 \\ 0 & 1 & 0 & -1 & 0 & 1 & 0 & 0 & 0 & 0 \\ 0 & 0 & 0 & 0 & 0 & 1 & 0 & 1 & 0 & 0 \\ 0 & 0 & 0 & 0 & 0 & 0 & 0 & -1 & -1 & 0 \end{bmatrix}$$

Fig. A.2 A directed graph and its loop incidence matrix, L

If all the elements of a spanning tree i.e. branches are grouped together and all elements representing chords are grouped separately, the loop incidence matrix can be written as

$$L = \begin{bmatrix} L_t & L_c \end{bmatrix} \qquad (A.3)$$

where L_t and L_c are loop incidence sub-matrices for branches and chords respectively. Size of sub-matrices L_t and L_c are $nl \times (nn-1)$ and $nl \times nl$. Notice that, since the nl columns of L_c correspond to chords defining the fundamental loops, L_c is a unit matrix.

One important relationship between node incidence matrix and loop incidence matrix is their orthogonality. The orthogonal relationship between N and L is written as

$$N L^T = 0 \qquad (A.4)$$

where superscript T stands for transpose of the matrix. Putting equations A.1 and A.3 in the equation A.4 yields

$$\begin{bmatrix} N_t & N_c \end{bmatrix} \begin{bmatrix} L_t & L_c \end{bmatrix}^T = 0 \qquad (A.5)$$

or

$$N_t L_t^T + N_c = 0 \qquad (A.6)$$

As N_t is a non-singular matrix, equation A.6 can be written as

$$L_t^T = -N_t^{-1} N_c \tag{A.7}$$

Equation A.7 relates sub-matrices of node incidence matrix and loop incidence matrix of a graph.

Appendix B | Results Concerning NLP Formulation of Analysis and Analysis-Design Problems

Results Concerning the Primal and Dual Formulations of Analysis Problem

(1) The Lagrangean function of Primal formulation of Analysis problem, given by eqs. 5.17-5.19, can be written as

$$L_P(q, \lambda_1, \lambda_2) = \sum_{i=0}^{neb} \int_0^{q_b} h_{b_i}(q_{b_i}) dq_{b_i} + \sum_{i=0}^{nea} \int_0^{q_a} h_{a_i}(q_{a_i}) dq_{a_i} + \sum_{i=0}^{nel} h_{1_i}^{*T} q_{1_i}$$

(B.1)

$$+ \lambda_1^T (q_1 - L_{21}^T q_2^* - L_{b1}^T q_b) + \lambda_2^T (q_a - L_{2a}^T q_2^* - L_{ba}^T q_b)$$

Differentiating with respect to q_1, q_a and q_b gives

$$\frac{\partial L_P}{\partial q_1} = h_1^* + \lambda_1 = 0 \quad \Rightarrow \quad \lambda_1 = -h_1^*$$

(B.2)

$$\frac{\partial L_P}{\partial q_a} = h_a + \lambda_2 = 0 \quad \Rightarrow \quad \lambda_2 = -h_a$$

(B.3)

$$\frac{\partial L_P}{\partial q_b} = h_b - L_{b1} \lambda_1 - L_{ba} \lambda_2 = 0$$

(B.4)

Note that Lagrange multipliers corresponds to the head loss in branches. Inserting the values of λ_1 and λ_2 from eqs. B.2 and B.3 into eq. B.4 gives loop head loss equation

$$h_b + L_{b1} h_1^* + L_{ba} h_a = 0$$

(B.5)

Differentiating eq. B.1 with respect to λ_1 and λ_2 gives continuity equations 5.18 and 5.19. This implies that any solution (q^*, h^*) of the analysis problem corresponds to an optimal solution of q^* of primal formulation. Note that h_2 can be evaluated directly from eq. 5.13.

Similarly, Lagrangean function of dual formulation, given by eqs. 5.22-5.24, can be written as

$$L_D(h,\lambda_1,\lambda_2) = \sum_{i=0}^{neb} \int_0^{h_{b_i}} q_{b_i}(h_{b_i})dh_{b_i} + \sum_{i=0}^{nea} \int_0^{h_{a_i}} q_{a_i}(h_{a_i})dh_{a_i} + \sum_{i=0}^{ne2} q_{2_i}^{*T} h_{2_i}$$

(B.6)

$$+ \lambda_1^T(h_2 + L_{21}h_1^* + L_{2a}h_a) + \lambda_2^T(h_b + L_{b1}h_1^* + L_{ba}h_a)$$

Differentiating with respect to h_2, h_b and h_a gives

$$\frac{\partial L_D}{\partial h_2} = q_2^* + \lambda_1 = 0 \quad \Rightarrow \quad \lambda_1 = -q_2^*$$

(B.7)

$$\frac{\partial L_D}{\partial h_b} = q_b + \lambda_2 = 0 \quad \Rightarrow \quad \lambda_2 = -q_b$$

(B.8)

$$\frac{\partial L_D}{\partial h_a} = q_a + L_{2a}^T \lambda_1 + L_{ba}^T \lambda_2 = 0$$

(B.9)

Note that Lagrange multipliers correspond to the flow in chords. Inserting values of λ_1 and λ_2 from eqs. B.7 and B.8 into eq. B.9 yields

$$q_a = L_{2a}^T q_2^* + L_{ba}^T q_b$$

(B.10)

Differentiation with respect to λ_1 and λ_2 gives loop head loss equations 5.23 and 5.24. This implies that any solution (q^*, h^*) of the analysis problem corresponds to an optimal solution of h^* of dual formulation. Note that q_1 can be evaluated directly from eq. 5.11.

(2) The objective functions $\phi(q_b)$ and $\psi(h_a)$ given by eqs. 5.15 and 5.22 respectively are strictly convex.

From equations 5.15 and 5.22, the following expressions of the Hessian matrices of ϕ and ψ are obtained by direct differentiation

$$\nabla^2\phi(q_b) = \frac{1}{2} \Delta q_b^T L_{ba}[h_a'] L_{ba}^T \Delta q_b$$

(B.11)

$$\nabla^2\psi(h_a) = \frac{1}{2} \Delta h_a^T L_{ba}^T [q_b'] L_{ba} \Delta h_a$$

(B.12)

where: h_a' is the diagonal matrix of head loss derivatives. The diagonal entries are the derivatives

$$h_{a_i}' = \frac{dh_{a_i}(q_{a_i})}{dq_{a_i}} \quad i = 1,\dots.nea$$

(B.13)

q_b' is the diagonal matrix of flow derivatives; the diagonal entries are the derivatives

$$q_{b_j}' = \frac{dq_{b_j}(h_{b_j})}{dh_{b_j}} \quad j = 1,\dots.neb$$

(B.14)

Since $h_a' > 0$ for all $i \in nea$ and $q_b' > 0$ for all $j \in neb$, it follows that the Hessian matrices are positive definite . This implies the strict convexity of ϕ and ψ.

Following on the same line as for analysis problem, similar results for NLP formulation of analysis-design problem can be obtained.

Appendix C | Algorithm for the Generation of Spanning Trees T_{13a} and T_{14a}

In this appendix, algorithm for the generation of spanning trees T_{13a} and T_{14a} in a network consisting of type 1, 2, 3, 4 and 5 elements is developed. Algorithm selects those type 5 elements as tree elements, which have maximum conveyance. In other words, algorithm generates spanning trees with its branches having maximum possible conveyance. Note that conveyance of some of the tree elements may be less than the conveyance of some chord elements. This is because if these chord elements with higher conveyance are selected as tree elements, generation of spanning trees T_{13a} and T_{14a} will not be possible.

Conveyance of an element type 5 is defined as its discharge carrying capacity. It is calculated in terms of discharge that can pass through it under unit head loss. Element, which can pass higher discharge under unit head loss has higher conveyance.

Algorithm for generating spanning trees T_{13a} and T_{14a}

Objective of this algorithm is to find a set of type 5 elements, called type a elements, which makes a spanning trees T_{13a} and T_{14a} in a network and has the maximum possible total conveyance.

Inputs to this algorithm are

1. Number of nodes, nn, number of elements, ne and node incidence matrix N of the network. Let this network and its node incidence matrix be called Network A and N_A respectively.

2. Conveyance capacities of type 5 elements.

Following are the steps of the algorithm for generating spanning trees T_{13a} and T_{14a} in network A.

1. Remove all type 2 and 4 elements and contract all type 1 and 3 elements in network A

After step 1, network consists of only type 5 elements. Let this network be called network B. The node incidence matrix of network B has the size nn-ne1-ne3 × ne-ne2-ne4. Let the node incidence matrix of network B be called N_B.

2. Remove all type 2 and type 3 elements and contract all type 1 and 4 elements in network A. Let this new network be called Network C. The node incidence matrix of this network may be defined as N_C. The node incidence matrix of network B has the same size as that of network B.

Both networks B and C have all type 5 elements of network A. Both the networks have same number of nodes also. However, both networks have different node incidence matrices. Let these two networks have nn' number of nodes and ne' number of elements, where nn' is equal to nn-ne1-ne3 and ne' is equal to ne-ne2-ne3. For defining a spanning tree in these two networks, nn'-1 number of elements are required.

3. Find all possible combinations of nn'-1 number of elements from the set of ne' elements. Number of possible combinations will be equal to $^{ne'}C_{nn'-1}$.

4. Knowing the values of conveyance of all type 5 elements, find the total conveyance of each combination or group of elements, which are obtained in step 3. Total conveyance of a group of elements may be taken as product of conveyances of all the elements making a group. Arrange these groups of elements in descending order of their total conveyances.

It is to mention here that not every group of elements will make a spanning tree in network B or network C and only some of the groups of elements will make spanning trees in both the networks.

5. Take first group of elements, which has the maximum total conveyance and check whether this set of elements provide a spanning tree in networks B. For checking whether the set of elements in the group makes a spanning tree in network B, consider all the elements in the group as type a elements and all other remaining elements as type b elements. Divide N_B in two parts as $N_B = [N_a \ N_b]$. Find the determinant of sub-matrix $\{N_a\}$. If the determinant is zero, the set of elements does not make a spanning tree in the network and if the determinant is non-zero, the set of elements makes a spanning tree in the network. If the set of elements in the first group does not make a spanning tree in network B, select next group of elements which has the highest total conveyance from the remaining groups of elements as obtained in step 4 and check whether this set of elements makes a spanning tree in network B. If the set of elements makes a spanning tree in network B i.e. if determinant $\{N_a\}$ is non-zero, check whether this set of elements makes a spanning tree in network C. For checking whether this set of elements makes a spanning tree in network C, same procedure shall be followed as carried out for network B. If this set of elements does not make a spanning tree in network C, select next group of elements which has the highest total conveyance from the remaining groups of elements as obtained in step 4 and check whether this set of elements makes a spanning tree in network B and network C.

The motive of step 5 is to obtain a group of elements, which has the maximum possible total conveyance and whose elements make spanning trees in both network B and C.

After step 5, a set of elements is obtained which makes a spanning tree in both networks B and C and this set of elements has the maximum total conveyance from among the sets whose elements can make spanning trees in networks B and C.

The set of type 5 elements so obtained are the branches or type a elements of network A and all other type 5 elements are chords or type b elements. Spanning tree T_{13a} will consist of all type 1, type 3 and type a elements and spanning tree T_{13a} will consist of all type 1, type 4 and type a elements.

Consider a network A as an example as shown in Fig. C1.

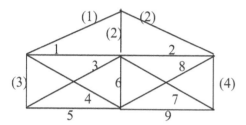

Numbers in bracket show element type
Numbers along edges show element no.
of type 5 elements

Fig. C1 : Network A as an example

After step 1 network B is obtained which is shown in Fig. C2.

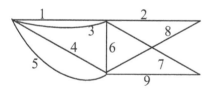

Fig. C2 : Network B after step 1.

After step 2 , network C is obtained, which is shown in Fig. C3.

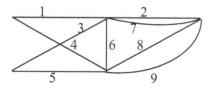

Fig. C3 : Network C after step 2.

Networks B and C have same number of type 5 elements and have same number of total nodes. There are 9 elements and 5 nodes, which means 4 number of elements will define a spanning tree in both networks. The total number of sets of 4 elements will be equal to $^9C_4 = 126$.

Let the conveyance of element no. 1 to 9 varies from 1 to 9. i.e. element no. 1 has conveyance 1 and element no. 9 has conveyance 9. Now these 126 sets are arranged in descending order of their total conveyance. The first set will consist of elements no. 6,7,8 and 9. The second set will consist of elements no. 5789 and the 126th set will consist of elements no. 1234.

The first set of elements is selected and it is checked whether this set of elements make a spanning tree in network B. As this set does not make a spanning tree in network B, next set of elements is selected i.e. set number 2 comprising of elements no. 5,7,8 and 9. As this set of elements also does not make a spanning tree in network B, next set is selected. The procedure is repeated till a set of elements is obtained which makes a spanning tree in both networks B and C. For the example network, this set comprises of elements no. 3,4,7 and 8.

The above algorithm is easy and can easily be implemented on computer.

References

Altman, T. and Boulos, P.F. (1995). "Solving flow constrained networks-inverse problem." *Journal of Hydraulics Division,* ASCE, Vol. 121, No. 5, May, 1995.

Bhave, P.R. (1990) "Rules for solvability of pipe networks." *Journal of Indian Water Works Association,* 22(1), 7-10, 1990.

Bhave, P.R. (1986). "Unknown pipe characteristics in Hardy-Cross method of network analysis." *Journal of Indian Water Works Association,* 18(2), 133-135, 1986.

Bhave, P.R. (1988). "Calibrating water distribution network models." *J. Envir. Engrg.,* ASCE, 114(1), 120-136, 1988.

Bhave, P.R. (1991). *Analysis of flow in water distribution networks.* Technomic Publishing Co., Inc., Lancaster, Pa. 1991.

Boulos, P.F., Altman, T. and Liou, J.C.P. (1993). "On the solvability of water distribution networks with unknown pipe characteristics." *Journal of Applied Mathematical Modelling,* Vol. 17, 1993.

Boulos, P.F., Altman, T. (1991). "A graph-theoretic approach to explicit nonlinear pipe network optimization." *Journal of Applied Mathematical Modelling,* Vol. 15, 1991.

Boulos, P.F., Altman, T. (1993). "An explicit approach for modelling closed pipes in water networks." *Journal of Applied Mathematical Modelling,* Vol. 17, 1993.

Boulos, P.F. and Wood, D.J. (1991). "An explicit algorithm for calculating operating parameters for water networks." *Journal of Civil Engineering Systems,* Vol. 8, 1991.

Boulos, P.F. and Wood, D.J. (1990). "Explicit calculation of pipe network parameters." *Journal of Hydraulic Division*, ASCE, 116(11), 1990.

Boulos, P.F. and Wood, D.J. (1991). "Explicit determination of network parameters for improving water distribution system performance." *2nd Int. Conference on Computer Methods and Water Resource,* Rabat, Morocco, 1991.

Boulos, P.F. and Wood, D.J. (1992). "Closure to explicit calculation of pipe network parameters." *Journal of Hydraulic Division,* ASCE, 118(7), 1992.

Chandarshekar, M. and Stewart, K.H. (1975). "Sparcity oriented analysis of large pipe networks." *Journal of Hydraulics Division,* ASCE, Vol. 101, No. HY4, April, 1975.

Chandarshekar, M. and Kesavan, H.K. (1972). "Graph-theoretic models for pipe network analysis." *Journal of Hydraulics Division,* ASCE, Vol. 98, No. HY2, Feb., 1972.

Chandarshekar, M. (1980). "Extended set of components in pipe networks." *Journal of Hydraulics Division,* ASCE, Vol. 106, No. HY1, Jan., 1980.

Chen, W. K. (1990). Theory of nets : flow in networks, Wiley, New York.

Collins, M.A., Cooper, L., Helgason, R.V. and Kenington, J.L. (1978). "Solution of large scale pipe networks by improved mathematical approaches." *Technical Report* IEOR 77016-WR-77001, School of Engineering and Applied Science, South Methodist University, Dallas, Texas, 1978.

Collins, M. A. (1980). "Pitfalls in pipe network analysis techniques". *J. Transp. Engg.,* ASCE, 106(5), 507-521.

Collins, M. , Cooper, L., Helgason, R., Kennington, J. L. and LeBlanc, L. (1978). "Solving the pipe network analysis problem using optimisation techniques." *Mgmt Sci.* 24(7), 747-760.

Coulbeck, B. and Sterling, M. (1978). "Optimal control of water distribution systems." *Proc. of Institute of Electrical Engineers*, 125(9), 1039-1044, 1978.

Coulbeck, B., Brdys, M., Orr, C. and Rance, J. (1988). "A hierarchial approach to optimized control of water distribution systems: Part I decomposition." *Journal of Optimal Control Applications and Methods,* 9(1), 51-61, 1988.

Coulbeck, B., Brdys, M., Orr, C. and Rance, J. (1988). "A hierarchial approach to optimized control of water distribution systems: Part II lower level algorithm." *Journal of Optimal Control Applications and Methods,* 9(2), 109-126, 1988.

Cross, H. (1936). "Analysis of flow in networks of conduits or conductor." *University of Illinois Engineering Experiment Station,* Bulletin No. 286, Champaign, 1936.

Demuren, A.O. (1986). "Pipe network analysis by partial pivoting method." *Journal of Hydraulics Division,* ASCE, Vol. 112, No. 5, May, 1986.

Epp, R. and Fowler, A. G. (1970). "Efficient code for steady state flows in networks". *J. Hydr. Div.*, ASCE, 96(1), 43-56.

Fletcher, R., (1987) Practical Methods of Optimization, John Wiley & Sons, New York.

Gessler, J. (1981). "Chapter 4: Analysis of pipe networks." *Closed conduit flow*, M. H. Choudhary and V. Yevjevich, eds., Water Resources Publications, Littleton, Colo.

Gill, P., Murray, E., and Wright, M., (1981). Practical Optimization. Academic Press. London.

Gofman, E. and Rodeh, M. (1981). "Loop equations with unknown pipe characteristics." *Journal of Hydraulics Division,* ASCE, 107(9), 1047-1060, 1981.

Goldberg, D. E., (1989) *Genetic Algorithms in Search, Optimization and Machine Learning.* Reading, MA: Addison-Wesley.

Gopal, M. (1989). *Modern Control System Theory*, Wiley Eastern Limited, New Delhi.

Gue, R. L. and Thomas, M. E., (1971) *Mathematical Methods in Operations Research*, The Macmillan, New York.

Gupta, R. K. (1997). "Transient Control by Valve Stroking in Pipe Networks". Fourth Convention Day on Design Related Research of Hydraulic Structures and Sanitary Engineering, Onderzoekschool Waterbouw, TU, Delft, The Netherlands.

Gupta, R. K. and Petry, B., (2004), "Use of Optimal Control Strategies : Rigid vs. Elastic Model", 9th *International Conference on Pressure Surges, BHR Group*, March   2004, Chester, UK

Holland, J., (1975). *Adaptation in Natural and Artificial Systems*. Ann Arbor, The University of Michigan Press.

Hutchinson, B. (1991). "Operational control of water distribution systems." *Proc. of Am. Water Works Assoc. Conf. on Comput. in the Water Industry*, American Water Works Association, 53-59, 1991.

Issacs, L.T. and Mills, K.G. (1980). "Linear theory methods for pipe network analysis." *Journal of Hydraulics Division,* ASCE, Vol. 106, No. HY7, July, 1980.

Issacs, L.T. and Mills, K.G. (1980). "Computer methods for pipe network analysis." *Queensland Division Technical Papers,* Institute of Engineers, Australia (Queensland Division), 21(10): 9-13, 1980.

Jowitt, P.W. and Xu, C. (1990). "Optimal valve control in water distribution networks." *J. Water Resour. Plng. and Mgmt.*, ASCE, 116(4), 1990.

Karney, B. W. (1990). " Energy relations in transient closed conduit flows". *J. Hydr. Engr.*, ASCE, 116(10), 1180-1196.

Karney, B. W. and Russ, E. (1985). "Charts for waterhammer in pipelines resulting from valve closure from full opening only." *Can. J. Civ. Engg., Ottawa*, 12(2), 241-264.

Lam, C.F. and Wolla, M.L. (1972). "Computer analysis of water distribution systems: Part I-Formulation of equations." *Journal of Hydraulics Division,* ASCE, Vol. 98, No. HY2, Feb., 1972.

Martin, D. W. and Peters, G. (1963). "The application of Newton's method to network analysis by digital computer." *J. Inst. of Water Engrs.*, 17, 115-129.

McInnis, D. and Karney, B. W. (1995). "Transients in distribution networks : field tests and demand models." *J. Hydr. Engg.*, ASCE, 121(3), 218-230.

Millar, W. (1951). "Some general theorems for non-linear systems possessing resistance." *Transactions,* IEEE, 1951.

Nielsen, H.B. (1989). "Methods of analyzing pipe networks." *Journal of Hydraulics Division,* ASCE, Vol. 115, No. 2, Feb., 1989.

O'Neill, I. C. and Graze, H. R. (1966). " Discussion on the rigid water column theory for uniform gate closure. By H. R. Valentine." *J. Hydr. Div.,* ASCE, 92(2), 382-387.

Onizuka, K. (1986). "System dynamics approach to pipe network analysis." *Journal of Hydraulics Division,* ASCE, Vol. 112, No. 8, Aug., 1986.

Ormsbee, L. E. and Wood, D. J. (1986a). "Explicit pipe network calibration". *J. Water Resour, Plng. And Mgmt.*, ASCE, 112(2), 166-182.

Ormsbee, L. E. and Wood, D. J. (1986b). "Hydraulic design algorithms for pipe networks". *J. Hydr. Engg.*, ASCE, 112(12), 1195-1207.

Parmakian, J. (1955). *Waterhammer analysis*. Prentice Hall, Englewood Cliffs, N.J.

Petry, B. and Gupta, R. K., (1998), "An Improved Approach to Surge Tank Design in High Head Hydropower Systems", *Proceedings of the International Conference on Modelling, Testing and Monitoring of Hydropower Plants – III*, Aix-En-Provence, France.

Petry, B and Gupta, R. K., (1999), "Valve Control of Pressure Surges in Hydraulic Systems", *XXVIII Congress of International Association for Hydraulic Research*, Graz, Austria.

Petry, B and Gupta, R. K., (1999), "Valve Operations to Control Pressure Surges in Pipe Networks", *Proceedings of the 3rd ASME/JSME Joint Fluids Engineering Conference*, San Francisco, USA.

Petry, B and Gupta, R. K., (2000), "Optimal Control of Pressure Surges in Pipe Networks", *8th International Conference on Pressure Surges, BHR Group*, April 2000, The Hague, The Netherlands.

Polak, E. (1971). *Computational methods in optimization : a unified approach.* Academic Press, New York.

Propson, T.P. (1970). "Valve stroking to control transient flows in liquid piping systems." *Ph.D Thesis,* University of Michigan, Ann Arbor, USA, 1970.

Ribeiro, C.R., Koelle, E. and Szajnbok, M. (1986). "Operational control of flow in hydraulic networks." *Proceedings of the 5th international conference on pressure surges,* Hannover, F.R.Germany, Sept., 1986.

Salgado, R., Todini, E. and O'Connell, P. E. (1988). "Extending the gradient method to include pressure regulating valves in pipe networks." *Proc. Int. Symp. On Comp. Modelling of Water Distribution Systems.* Kentucky Water Resour. Res. Inst. and College of Engg., Uni. Of Kentucky, 157-180.

Shamir, U. and Howard, C.D.D. (1968). "Water distribution system analysis." *Journal of Hydraulics Division,* ASCE, 94(1), 219-234, 1968.

Shamir, U. and Howard, C.D.D. (1970). "Closure to water distribution system analysis." *Journal of Hydraulics Division,* ASCE, 96(2), 577-578, 1970.

Shamir, U. and Howard, C.D.D. (1977). "Engineering analysis of water distribution systems." *Journal of American Water Works Association,* 69(9),510-514, 1977.

Shimada, M. (1992). "State-space analysis and control of slow transients in pipes." *Journal of Hydraulics Division,* ASCE, Vol. 118, No. 9, Sept., 1992.

Shimada, M. (1988). "Time marching approach for pipe steady flows." *Journal of Hydraulics Division,* ASCE, Vol. 114, No. 11, Nov., 1988.

Stephenson, D. (1966). "water hammer charts including fluid friction." *J. Hydr. Div.,* ASCE, 92(5).

Stoner, M.A. (1968). "Analysis and control of unsteady flows in natural gas piping systems." *PhD Thesis,* University of Michigan, Ann Arbor, USA, 1968.

Streeter, V.L. (1963). "Valve stroking to control water hammer." *Journal of Hydraulics Division,* ASCE, Vol. 89, No. HY2, Mar., 1963.

Streeter, V.L. (1967). "Valve stroking for complex piping systems." *Journal of Hydraulics Division,* ASCE, Vol. 93, No. HY3, May, 1967.

Todini, E. and Pilati, C. (1988). "A gradient algorithm for the analysis of pipe networks." *Proc. Comp. Applications in Water Supply*, B. Coulbeck and C. Orr, eds., Research Studies Press, Letchworth, Hertfordshire, England, 1-20.

Valentine, H. R. (1965). "Rigid water column theory for uniform gate closure." *J. Hydr. Div.*, ASCE, 91(4), 27-33.

Watters, G. Z. (1984). *Analysis and control of unsteady flow in pipelines*. Butterworth Publishers, Stoneham, Mass.

Wood, F. M. (1973). "Comparision of the rigid column and the elasrtic theories for water hammer surges." *Proc. 1st Can. Hydr. Conf.*, 564-578.

Wood, D.J. and Charles, C.O.A. (1972). "Hydraulic network analysis using linear theory." *Journal of Hydraulics Division,* ASCE, Vol. 98, No. HY7, July, 1972.

Wood, D. J. and Rayes, A. G. (1981). " Reliability of algorithms for pipe network analysis. " *J. Hydr. Div.*, ASCE, 107(10), 1145-1161.

Wylie, E. B., and Streeter, V. L. (1993). *Fluid transients in systems*. Prentice Hall, Englewood Cliffs, N.J.

Zessler, U. and Shamir, U. (1989). "Optimal operation of water distribution systems." *Journal of Water Resource Planning and Management*, ASCE, 115(6), 1989.

List of Publications Arising Out of Present Study

1. Gupta, R. K. (1997). "Transient Control by Valve Stroking in Pipe Networks". Fourth Convention Day on Design Related Research of Hydraulic Structures and Sanitary Engineering, Onderzoekschool Waterbouw, TU, Delft, The Netherlands.

2. Gupta, R. K. and Petry, B., (2004), "Use of Optimal Control Strategies : Rigid vs. Elastic Model", 9th *International Conference on Pressure Surges, BHR Group*, March 2004, Chester, UK

3. Petry, B. and Gupta, R. K., (1998), "An Improved Approach to Surge Tank Design in High Head Hydropower Systems", *Proceedings of the International Conference on Modelling, Testing and Monitoring of Hydropower Plants – III*, Aix-En-Provence, France.

4. Petry, B and Gupta, R. K., (1999), "Valve Control of Pressure Surges in Hydraulic Systems", *XXVIII Congress of International Association for Hydraulic Research*, Graz, Austria.

5. Petry, B and Gupta, R. K., (1999), "Valve Operations to Control Pressure Surges in Pipe Networks", *Proceedings of the 3rd ASME/JSME Joint Fluids Engineering Conference*, San Francisco, USA.

6. Petry, B and Gupta, R. K., (2000), "Optimal Control of Pressure Surges in Pipe Networks", *8th International Conference on Pressure Surges, BHR Group*, April 2000, The Hague, The Netherlands.

7. Gupta, R. K. and Petry, B., "An Improved Approach to Network Analysis and Design, Part I – Solvability rules". To be submitted to *J. Hydr. Engr.*, ASCE.

8. Gupta, R. K. and Petry, B., "An Improved Approach to Network Analysis and Design, Part II – Algorithm for Solution". To be submitted to *J. Hydr. Engr.*, ASCE.

Printed and bound by CPI Group (UK) Ltd, Croydon, CR0 4YY

01/11/2024

01782610-0002